Climate Change and the Oceans

Climate Change and the Oceans

Gauging the Legal and Policy Currents in the Asia Pacific and Beyond

Edited by

Robin Warner

Australian National Centre for Ocean Resources and Security (ANCORS), University of Wollongong, Australia

and

Clive Schofield

Australian National Centre for Ocean Resources and Security (ANCORS), University of Wollongong, Australia

Edward Elgar

Cheltenham, UK • Northampton, MA, USA

Published by
Edward Elgar Publishing Limited
The Lypiatts
15 Lansdown Road
Cheltenham
Glos GL50 2JA
UK

Edward Elgar Publishing, Inc.
William Pratt House
9 Dewey Court
Northampton
Massachusetts 01060
USA

A catalogue record for this book
is available from the British Library

Library of Congress Control Number: 2012943182

MIX
Paper from
responsible sources
FSC® C018575
FSC
www.fsc.org

ISBN 978 1 84844 818 6

Typeset by Servis Filmsetting Ltd, Stockport, Cheshire
Printed and bound by MPG Books Group, UK

Contents

Figures

Tables

Contributors

Andi Arsana is a Lecturer and Researcher in the Department of Geodetic Engineering, Gadjah Mada University, Indonesia. He is currently an Australian Leadership Awards Scholar (PhD candidatc) at the Australian National Centre for Ocean Resources and Security (ANCORS), University of Wollongong. He obtained a Masters degree from the University of New South Wales, Australia, focusing on maritime delimitation between Indonesia and Timor L'Este. Andi is also an alumnus of the UN-Nippon Foundation Fellowship, a research and internship program organized by the UN Division for Ocean Affairs and the Law of the Sea (DOALOS). His research focuses on technical and geodetic aspects of the law of the sea, especially on maritime boundaries. He has published several works in the field including journals, books, and popular-scientific articles.

Marcus Haward is Associate Professor at the Institute for Marine and Antarctic Studies, University of Tasmania. His current research focuses on Antarctic and oceans governance, ecosystems aspects of fisheries management, knowledge systems in relation to coastal management, and climate adaptation. He has published extensively in these areas. Recent books include *Oceans Governance in the Twenty-first Century: Managing the Blue Planet*, with Joanna Vince (Edward Elgar, 2008) and *Australia and the Antarctic Treaty System*, co-editor with Tom Griffiths (NSW Press, 2011).

Ove Hoegh-Guldberg is the Director of the Global Change Institute (GCI) and is Professor of Marine Science at the University of Queensland in Brisbane, Australia. Ove has a BSc (honours) from Sydney and a PhD from the University of California, Los Angeles. His research focuses on the impacts of global warming and ocean acidification on coral reef ecosystems. He has held academic positions at the University of California, Los Angeles; Stanford University; the University of Sydney; and the University of Queensland. He is a member of Climate Scientists Australia; the Royal Society (London) Marine Advisory Network; and the Board of Editing Reviewers at *Science* magazine, and is the Coordinating Lead Author for the 'Oceans' chapter for the fifth assessment report of the Intergovernmental Panel on Climate Change. In 1999 he was awarded the

Eureka Prize for his scientific research. He is currently the Queensland Smart State Premier's Fellow (2008–13) and an Australian Research Council Laureate Fellow.

Stuart Kaye is Dean and Winthrop Professor of Law at the University of Western Australia. He holds degrees in arts and law from the University of Sydney, winning the Law Graduates' Association Medal, and a doctorate from Dalhousie University. He has written a number of books, including *Australia's Maritime Boundaries*, *The Torres Strait*, and *International Fisheries Management*. He was appointed to the List of Arbitrators under the Environmental Protocol to the Antarctic Treaty in 2000 and was elected a Fellow of the Royal Geographical Society in 2007.

Richard Kenchington is a Professorial Fellow at the Australian National Centre for Ocean Resources and Security (ANCORS). He is currently leader of the Integration, Analysis and Synthesis Theme for a CSIRO Coastal Cluster research project addressing barriers to the uptake of science in coastal zone management. A marine ecologist, he studied crown of thorns starfish and large-scale surveys of the Great Barrier Reef. Later he was a foundation member of the staff of the Great Barrier Reef Marine Park Authority in 1978, retiring in 1999 as Executive Director of the Authority. He has published on ecosystem based management of coral reef, coastal and marine ecosystems. He works widely as a consultant and advisor to governments and international agencies.

Lisa Kirkendale studied at the University of Victoria (UVic)/Bamfield Marine Sciences Centre (British Columbia, Canada), University of Guam Marine Laboratory (Mangilao, Guam) and the University of Florida (Gainesville, USA). She was a post-doctoral research associate with School of Earth and Ocean Sciences (UVic) prior to taking up the post of Curator of Invertebrates at the Royal British Columbia Museum in Victoria, Canada. She recently relocated to Australia and is now a Research Fellow with the Shoalhaven Marine and Freshwater Centre, University of Wollongong, where her research aim is to infuse applied marine science with an evolutionary twist.

Stathis Palassis is a Senior Lecturer at the Faculty of Law at the University of Technology Sydney. He holds a Bachelor of Jurisprudence/Bachelor of Laws from the University of Western Australia and Doctor of Juridical Studies from the University of Sydney; and prior to entering academia he was a litigation lawyer in New South Wales. His primary area of research is on the Law of the Sea/International Environmental Law, and in particular on environmental aspects concerning the international shipping sector including: environmental regulation of international shipping; liability and compensation for ship-sourced environmental damage; climate change and shipping; and shipping and the polar regions.

Chris Rahman is Senior Research Fellow in Maritime Strategy and Security at the Australian National Centre for Ocean Resources and Security (ANCORS), University of Wollongong. He is an academic strategist researching contemporary issues in maritime strategy, Australian defence policy, China, and the strategic relations of the Indo-Pacific region. In addition, he is coordinator of major projects on the maritime security implications of climate change and the history of the Pacific Patrol Boat Project; and manages the ANCORS Vessel Tracking Initiative in collaboration with industry and Australian government partners.

Rosemary Rayfuse holds positions as Professor of International Law in the Faculty of Law, University of New South Wales, Sydney, Australia and Conjoint Professor of International Environmental Law in the Faculty of Law at Lund University in Sweden. Her publications include *Non-Flag State Enforcement in High Seas Fisheries* (Martinus Nijhoff, 2003) and (with Shirley Scott) *International Law in the Era of Climate Change* (Edward Elgar, 2012), as well as numerous articles on issues related to fisheries enforcement, oceans governance and protection of the marine environment in areas beyond national jurisdiction. Her current research focuses on the normative effects of climate change on international law.

Ana Rubio is a marine ecology and aquaculture scientist at the Shoalhaven Marine and Freshwater Centre at the University of Wollongong. She undertook an MSc in Fisheries and Aquaculture at the University of Wales, Bangor (UK) and obtained her PhD degree at the Australian National University. Her work combines environment, health of catchments and waterways and aquaculture industry practices. Recently she has been working with the Australian oyster industry assisting them with the identification of catchment risks and improvement of their husbandry practices as part of the implementation of estuary-wide Environmental Management Studies, as well as developing catchment-wide data portals for the better management of oyster areas and undertaking research on environmental controls of oyster farms and new technologies to improve industry sustainability.

Clive Schofield is Professor and Director of Research at the Australian Centre for Ocean Resources and Security (ANCORS), University of Wollongong, Australia. He currently holds an Australian Research Council Future Fellowship. He is a political geographer and international legal scholar whose research interests relate to international boundaries, and particularly maritime boundary delimitation. He holds a BA (Hons) and PhD in Geography from the University of Durham in the UK (1990 and 2000), and an LLM (with distinction) from the Faculty of Law at the University of British Columbia (UBC) in Canada (2009). He is co-author, with Professor Victor Prescott, Emeritus Professor of Geography, University of Melbourne, of the book, *The Maritime Political Boundaries of the World* (2005). He has been involved in the peaceful settlement of boundary and territory disputes, for example in multiple cases before the International Court of Justice.

Robin Warner is Associate Professor at the Australian National Centre for Ocean Resources and Security, University of Wollongong. She holds a PhD from the University of Sydney, a Master of International Law from the Australian National University and LLB and BA (Hons) degrees from the University of Sydney. Her current research interests include law of the sea, oceans governance, marine environmental law, climate law and transnational criminal law. She is the author of *Protecting the Oceans Beyond National Jurisdiction: Strengthening the International Law Framework* (Martinus Nijhoff, Leiden, 2009) and editor (with Simon Marsden) of *Transboundary Environmental Governance: Inland Coastal and Marine Perspectives* (Ashgate Publishers, Farnham, UK, 2012), as well as many book chapters and journal articles on oceans law and policy.

Pia C. Winberg is currently the Director of the Shoalhaven Marine & Freshwater Centre at the University of Wollongong. Pia's background is in marine systems ecology and her main research interest is in marine food production systems that are sustainably integrated with the coastal and marine environment. She has undertaken research and development with aquaculture industries in temperate Australia and in Sri Lanka, including for fish, prawns, abalone and oysters, and has developed pilot-scale, trophically-integrated marine aquaculture systems with private industry. She has also been a key driver of Seaweeds Australia, focusing on the development of high value seaweed industries.

Abbreviations

AAT	Australian Antarctic Territory
AF	Assessment Framework
ASAC	Antarctic Science Advisory Committee
ATCM	Antarctic Treaty Consultative Meetings
ATCP	Antarctic Treaty Consultative Parties
ATS	Antarctic Treaty System
CAML	Census of Antarctic Marine Life
CBD	Convention on Biological Diversity
CBDR	common but differentiated responsibilities
CCAMLR	Convention for the Conservation of Antarctic Marine Living Resources
CCS	carbon capture and storage
CDM	Clean Development Mechanism
CER	certified emission reduction
CFC	chlorofluorocarbons
CITES	Convention on International Trade of Endangered Species of Wild Fauna and Flora
CLCS	Commission on the Limits of the Continental Shelf
CMS	Convention on Migratory Species of Wild Animals
COAG	Council of Australian Governments
COP	Conference of the Parties
CTI	Coral Triangle Initiative
DIC	dissolved inorganic carbon
EBSA	ecologically and biologically significant areas
ECA	Emission Control Areas
EECA	Energy Efficiency and Conservation Authority
EEDI	Energy Efficiency Design Index
EEOI	Energy Efficiency Operational Indicator
EEZ	exclusive economic zone
EIA	environmental impact assessment
ENSO	El Niño-Southern Oscillation
EOR	enhanced oil recovery
EPBC	Environment Protection and Biodiversity Conservation

ETS	emissions trading scheme
EWTEC	European Wave and Tidal Energy Conference
FAO	Food and Agriculture Organization
FERC	Federal Energy Regulatory Commission
GCM	Global Circulation Models
GDP	gross domestic product
GEF	Global Environment Facility
GHG	greenhouse gases
GT	gross tonnage
GW	gigawatts
HNLC	high-nutrient low-chlorophyll
IAPPC	International Air Pollution Prevention Certificate
ICAO	International Civil Aviation Organization
ICF	International Compensation Fund
ICOE	International Conference on Ocean Energy
ICRI	International Coral Reef Initiative
ICRW	International Convention for the Regulation of Whaling
ICSU	International Council for Science
IEA	International Energy Agency
IGY	International Geophysical Year
IMO	International Maritime Organization
IOC/UNESCO	Intergovernmental Oceanographic Commission of the United Nations Educational, Scientific and Cultural Organization
IPCC	Intergovernmental Panel on Climate Change
IPY	International Polar Year
ISPS	International Ship and Port Facility Security
IUCN	International Union for Conservation of Nature
JI	joint implementation
LC	London Convention – 1972 Convention on the Prevention of Marine Pollution by Dumping of Wastes and other Matter
LDC	least-developed countries
LEED	Low Emissions Energy Development
LIS	Leveraged Incentive Scheme
LNLC	low-nutrient low-chlorophyll
LOSC	Law of the Sea Convention
LP	London Protocol – 1996 Protocol to the London Convention
MBM	market-based measures
MEPC	Marine Environment Protection Committee

METS	Maritime Emissions Trading Scheme
MMS	Mineral Management Service
MOU	Memorandum of Understanding
MSC	Maritime Safety Committee
NOAA	National Oceanic and Atmospheric Administration
nm	nautical miles
NSCABD	The National Strategy for the Conservation of Australia's Biodiversity
OCCS	offshore carbon capture and storage
ODS	ozone-depleting substances
PACC	Pacific Adaptation to Climate Change
PCB	polychlorinated biphenyls
PEMSEA	Partnerships in Environmental Management for the Seas of East Asia
PSI	Proliferation Security Initiative
REZ	Renewable Energy Zone
RM	Rebate Mechanism
SARP	Standards and Recommended Practices
SEEMP	Ship Energy Efficiency Management Plan
SIDS	small-island developing states
SOIREE	Southern Ocean Iron Release Experiment
SOLAS	International Convention for Safety of Life at Sea
SPREP	South Pacific Regional Marine Environment Programme
SRES	Special Report on Emission Scenarios
SUA	Safety of Maritime Navigation Convention
TSS	traffic separation scheme
UK	United Kingdom
UN	United Nations
UNCED	United Nations Conference on Environment and Development
UNDP	United Nations Development Programme
UNFCCC	United Nations Framework Convention on Climate Change
US	United States
VES	Vessel Efficiency System
VLCT	very large commercial transporters
WMO	World Meteorological Organization
WTO	World Trade Organization

Foreword

Climate change is, undoubtedly, one of the greatest challenges of our time. Broad consensus has emerged in the global scientific community that global warming, occasioned by excess greenhouse gases in the atmosphere resulting from human activities, poses serious threats to the very survival of our planetary system, not least our oceans. Paradoxically, however, it is also now recognized that the oceans provide a partial solution to the global climate change problem as a result of their role in the carbon cycle, ocean fertilization to mitigate impacts of climate change, providing alternative renewable energy sources, and as a reservoir for carbon sequestration and storage. These mitigation opportunities are likely to have significant negative consequences unless properly managed.

At the broader oceans governance level, the current rules of international law, codified largely in the 1982 United Nations Convention on the Law of the Sea, assign oceans governance responsibilities to States and the international community as a whole, on the basis of well defined maritime zones of jurisdiction. However, the specific rules laid down for the determination of maritime jurisdictional space did not take into account the climate change phenomenon and its likely impact on oceans jurisdictions, particularly in terms of the determination of baselines, the demarcation of maritime boundaries and in extreme situations, the inundation or total disappearance of islands.

The strong nexus between climate change impacts on the oceans and the mitigation role the oceans can play in the global response to the climate change threat demand a new governance framework at the international, regional and national levels, an aspect which is currently not well articulated in the proliferation of literature on climate change. This collection of essays, presented as *Climate Change and the Oceans: Gauging the Legal and Policy Currents in the Asia Pacific Region and Beyond* has done what has hitherto not been attempted. The issues presented in the various chapters are anticipatory, bold and relevant to the contemporary discourse on climate change. The book not only summarizes the issues cogently and comprehensively, but also presents them in a style which is easily understood. Although the title of the book is focused on the Asia Pacific region, the broader issues canvassed have relevance well beyond this region.

From the policy and governance perspectives, there are three key messages in *Climate Change and the Oceans*. The first is that addressing climate change's impacts on the oceans presents new opportunities for global cooperation which should be fostered. The second is the need to develop a global framework to monitor and regulate marine geo-engineering activities in response to climate change mitigation efforts. Finally, there is the crucial need to develop coherent international, regional and national policy responses and frameworks that recognize climate change impacts on the oceans. To be effective, this policy framework must be based on a proper understanding of the interlocked nature of the issues. There is also an urgent need to align national policies with regional and global initiatives.

Martin Tsamenyi,
Professor of Law & Director, Australian National Centre for Oceans Resources & Security,
University of Wollongong, Australia

1. Climate change and the oceans: legal and policy portents for the Asia Pacific region and beyond

Robin Warner and Clive Schofield

How inappropriate to call this planet Earth when it is quite clearly Ocean.
(Arthur C. Clarke, *Nature*, v. 344, p. 102)

1. INTRODUCTION

The oceans dominate the globe spatially, covering approximately 72 per cent of its surface area. These extensive marine spaces are critical to the global environment and human survival in numerous ways – they are vital to the global nutrient cycling, represent a key repository and supporter of biological diversity on a world scale, and play a fundamental role in driving the global atmospheric system. Moreover, the oceans continue to provide a critical source of food through fisheries and aquaculture, are an increasingly significant source of energy resources, and underpin the global economy through sea-borne trade.

The oceans are, however, under increasing threat. In addition to well recognized threats such as pollution and dumping, as well as urgent concerns over over-fishing, the destruction of valuable habitats and the preservation of marine biodiversity, the oceans have had a longstanding role as the primary sink for excess heat and carbon present in the global climate system. Consequently, the oceans have and are continuing to experience profound impacts as a result of climate change. In particular, substantial impacts on marine environments will result from increasing water temperatures, changes to the chemistry of seawater, including ocean acidification, changes in ocean circulation systems leading to shifts in the strength and direction of ocean currents and increases in the geographical range, frequency and intensity of extreme weather events. Potentially substantial and abrupt sea-level rise is also likely to have dramatic consequences.

The multidimensional impacts of climate change are likely to pose disastrous and destabilizing threats for many maritime regions with potentially dire consequences for marine and coastal environments, resources and populations. Inspired by mutual interest in one of the most important challenges of our times, this book has risen from ongoing research by the contributing editors Robin Warner and Clive Schofield into the legal and policy implications of climate change for the world's oceans. The volume draws together a range of informed opinions by leading scholars from diverse disciplines on the impacts of climate change on the oceans with a particular emphasis on the Asia Pacific region (taken to mean the broad geographical area bordering the western Pacific Ocean including East and Southeast Asia, the Pacific Island States and Oceania), but also taking into account global developments. Legal and policy responses to those developments are explored at the global, regional and national levels, and from the perspectives of different maritime sectors. The book aims to raise awareness of the intensifying effects of climate change on the oceans and the need for proactive and balanced governance of these to mitigate adverse impacts, both globally and within the Asia Pacific region.

This introductory chapter seeks to draw out some of the key themes explored in the volume. These are all associated with the actual and potential impacts of climate change on the oceans, notably in terms of the marine environment, living resources such as fisheries, and marine biodiversity. Anticipated impacts on the scope of maritime jurisdictional claims on the part of coastal States as a result of sea level rise are outlined, as well as the potential repercussions for contentious maritime and territorial disputes are explored. The implications of such developments for maritime security present a further focus, before attention shifts to options to mitigate the impacts of climate change on the oceans.

2. CLIMATE CHANGE IMPACTS ON THE OCEANS

As the oceans absorb more of the heat generated by global warming, seawater expands and causes a rise in sea levels. Even discounting the effects of glacial and ice sheet melting, the Intergovernmental Panel on Climate Change (IPCC) has projected a global average rise in sea levels of 30–40 cm by the end of the century (IPCC 2007). It has also forecasted that a changing climate, including alterations to wind patterns and storm climatology, are likely to produce more intense tropical storms, which, in combination with sea level rise, will result in more frequent storm surges and thus more severe flooding (IPCC 2007). This will have varying effects in the Asia Pacific region depending primarily on coastal topography. At particular risk are river deltas such as those in the South East Asian region, which over the long term will be permanently submerged or eroded away. For example, Vietnam, which possesses the major deltas of the Red River and, especially, the Mekong River, has been predicted to lose over 14,000 square miles of delta areas, where substantial portions of its population, as well as some of its most productive agricultural land, are concentrated. Significant delta areas are similarly disappearing in Thailand, Burma and Indonesia (CENTRA 2010; Jasparro and Taylor 2008). The combined effects of sea-level rise, increased storm activity and changes in sea temperature and salinity are likely, for example to further intensify already existing threats to important coastal environments such as the mangrove forests of the region, and coastal deltas in South East Asia. These areas provide crucial natural protection for the rest of the coast and their loss is likely to lead to substantial erosion. Additionally, mangroves offer important habitats that currently play a vital role in biodiversity preservation and, for instance, in sustaining regional fisheries upon which millions depend as their primary source of protein. Sea-level rise, flooding and storm activity may also lead to the inundation of multiple small and, critically, low-lying islands in the region, especially in the Indonesian and Philippine archipelagos, with Indonesia projected to lose 2000 small islands by 2030 (CENTRA 2010). The threatened retreat of shorelines also poses potential knock-on effects in terms of impacts on baselines and thus the spatial scope of national claims to maritime jurisdiction. Coastal cities, towns and villages in the region will also confront the risk of inundation from sea level rise and wild weather resulting in the destruction of housing and infrastructure. Indeed, even if low-lying islands and mainland coastal areas are not entirely submerged, repeated flooding coupled with salt-water intrusion is likely to have negative implications for the continued habitability of these areas. As well as socioeconomic disruption and potential civil unrest within States of the region, the aftermath of sea-level rise and the severe weather events associated with climate change may also lead to mass displacements of people seeking shelter and economic viability beyond their own States (Barnett and Adger 2007; Smith 2007). This would add to the already high numbers of asylum seekers moving among islands in South East Asia and to proximate countries such as Australia and New Zealand.

The rising temperature of global oceans will result in the relocation and in some cases extinction of marine species. The 2008 *Report of the FAO Expert Workshop on Climate Change Implications for Fisheries and Aquaculture* highlights the nature and extent of climate change impacts expected in the South China Sea and adjacent South East Asian and Pacific waters. As noted above these include higher water temperatures; sea-level rise; an increase in the frequency and intensity of storms; changes in the quantity, location and timing of rainfall; as well as changes in the effects of the El Niño Southern Oscillations (Food and Agriculture Organization [FAO] 2008). In particular, an increase of 0.6–0.7° in water temperature is predicted for the tropical regions including the South China Sea and South East Asia, leading to increased stratification and reduced mixing of water in lakes, reduced primary productivity and ultimately reduced food supplies for fish species (WorldFish Center 2006). Higher water temperatures will result in more frequent algal blooms, less dissolved oxygen, and an increased incidence of disease and parasites, which, in turn, will lead to less abundant species and composition in fish stocks, as well as substantial increases in infrastructure and operating costs for aquaculture enterprises. Higher water temperatures are also forecast to affect the timing and success of fish migrations, spawning, sex ratios and peak abundance for some species (WorldFish Center 2006). The damage and destruction to coral reefs and other marine ecosystems associated with increased ocean acidity is also likely to erode the economic security of many States in the Asia Pacific region by removing vital fisheries habitats and sources of tourist industry income. A specific example of the effects of higher water temperatures on fish abundance is the anticipated decline in skipjack tuna in the waters of the Coral Triangle, directly to the south and east of the South China Sea large marine ecosystem (Willis et al. 2008).

As coral reefs and other fragile marine ecosystems degrade, processes such as environmental impact assessment and marine spatial planning, including the designation of marine protected areas in coastal zones and beyond, will become increasingly important in conserving marine living resources and biodiversity. The adverse effects of climate change on the oceans also highlight the need for coastal States, regional fisheries management organizations and regional seas organizations to adjust their conservation and management regimes in response to a changing and dwindling pool of resources and increasing threats to biodiversity. This book examines the potential patterns and locations of climate change effects on marine biodiversity in the Asia Pacific region and analyses current and projected legal and policy responses to these dimensions of climate change.

Continuing sea level rise, which is already affecting the South China Sea and South East Asia, is expected to have similarly deleterious effects on fisheries and aquaculture productivity. As noted above, Vietnam is especially vulnerable and has already experienced a 2.5–3 cm increase in sea level over the last 50 years. In the Lower Mekong Basin the sea level has risen by as much as 3 cm, with further increases predicted, possibly by as much as 1 m by the end of the century (Mekong River Commission [MRC] 2010; see also Doyle et al. 2010). The loss of land and the saltwater intrusion induced by sea level rise has resulted in damage to and loss of freshwater fisheries, and a reduced area available for land-based aquaculture (WorldFish Center 2006). The increased likelihood of severe weather events associated with climate change such as large waves and storm surges poses higher direct risks to fishers, including loss of equipment and aquaculture stock and damage to aquaculture facilities (WorldFish Center 2006). Reponses to salinity changes have prompted the building of more dykes and water gates along coastlines in the South East Asian region, fragmenting fisheries habitats and disturbing fish migration and spawning (WorldFish Center 2006; CENTRA 2010).

The changed location and timing of ocean currents and upwelling linked to climate change is also expected to alter the nutrient supply to fish in surface waters, leading to changes in distribution and productivity of open sea fisheries (WorldFish Center 2006). A further well-documented effect of climate change and land-based pollution is the bleaching of coral reefs, resulting in reduced reef fish productivity (CENTRA 2010).

3. CLIMATE CHANGE AND MARITIME JURISDICTION

Rising sea levels also have the potential to significantly affect the extent of coastal State claims to maritime jurisdictional zones and the valuable marine resources within them. This threat of the loss of currently claimed maritime spaces arises because of the fundamental link between the limits of maritime zones and the baselines along the coast from which such zones are measured. Most baselines are 'normal' ones coincident with the low-water line along the coast which, in turn, is vulnerable to retreat inland in the face of sea level rise. As baselines are pulled back inland, so the limits of maritime jurisdiction measured from them will likewise shift position, leading to a reduction in maritime spaces under the jurisdiction of the coastal State concerned. This problem persists even when straight or archipelagic baselines have been defined, as is the case for a number of States in the Asia Pacific, because such straight-line type baselines still need to be tied back and anchored to normal, low-water line baselines which are susceptible to change in the face of sea-level rise. Similarly, sea-level rise has the potential to lead to the inundation of small islands or, alternatively, their depopulation and change in status, with significant potential reductions in the capacity of these features to generate extensive maritime claims. Significant reductions in the scope of coastal State maritime claims inevitably has implications in terms of jurisdiction over and access to valuable marine resources, both living, such as fisheries, and non-living, including seabed hydrocarbons.

As the effects of climate change intensify, and impacts on baselines and maritime jurisdictional limits increasingly come to the fore, significant repercussions for the negotiation of maritime boundaries and the resolution of disputes may arise. While boundary treaties have a privileged position in international law with a view to ensuring their permanence and inviolability, nonetheless as coastlines recede due to erosion and flooding caused by sea-level rise there may be calls to reopen apparently settled boundary disputes. Moreover, the majority of potential maritime boundaries globally have yet to be settled and the Asia Pacific region features multiple notable territorial and maritime disputes of long standing (Prescott and Schofield 2005). These include disputes concerning sovereignty over small islands in the East China Sea between China and Japan and multiple groups of islands in the South China Sea, including the infamous dispute over the Spratly Islands, which involves six claimants (Brunei, China/Taiwan, Malaysia, the Philippines and Vietnam). In large part though not wholly as a result of the sovereignty disputes over these small features, these semi-enclosed seas both feature broad areas of overlapping maritime claims and serious maritime boundary disputes. Additionally, notable maritime jurisdictional disputes exist in the Gulf of Thailand, particularly between Cambodia and Thailand and also in the Sulawesi (Celebes) Sea, involving Indonesia and Malaysia.

The impacts of sea-level rise on baselines, islands and maritime claims may well lead to fluctuations in the positions taken by States in maritime delimitation negotiations, complicating settlement of existing disputes as fresh arguments emerge over which baselines should be used in the construction of equidistance lines which are the frequent starting points for discussions (Caron 1990; Prescott and Schofield 2005; Paskal 2007). The inundation of islands will also potentially affect sovereignty claims over them, as well as their sovereign rights to offshore maritime resources (Freestone 1990; Schofield 2009). For example, with the onset of sea-level rise many of the Spratly Islands group may be permanently submerged, potentially undermining the basis for territorial sovereignty claims based on these insular features. However, States rarely concede sovereignty over territory easily, and such disputes may persist should the claimant and occupying States undertake concerted reclamation or protection works in order to preserve the threatened, yet disputed, features. Such actions would themselves be likely to generate further friction among the disputants. That said, the Asia Pacific region is also host to innovative practice in respect of provisional arrangements of a practical nature designed to address the challenge of contentious maritime disputes and overlapping maritime claims. Examples include maritime joint development arrangements designed to facilitate exploration for seabed hydrocarbons in the Gulf of Thailand between Malaysia and Thailand and Malaysia and Vietnam, as well as joint fishing zones between China and Vietnam in the Gulf of Tonkin (Bac Bo Gulf in Vietnamese, Beibu Gulf in Chinese) in the northern South China Sea, and several joint fishing arrangements in the East China Sea (between China and Japan, China and Korea and Japan and Korea). Perhaps more urgently, as the adverse impacts of climate change deepen for many of the States of the Asia Pacific involved in maritime disputes, especially those bordering the East China Sea and South China Sea, heightened competition for dwindling resources will increase the potential for a re-escalation of tensions (Schofield et al. 2011). If claimant States choose to enforce their claims using military power this could lead to more confrontations at sea (Bradford 2005).

4. CLIMATE CHANGE AND MARITIME SECURITY

Sea-level rise coupled with more severe weather events centring in the ocean – including storm surges – may threaten the territory of coastal States and lead to small island States becoming uninhabitable. These developments are likely to have wide-ranging security implications for the Asia Pacific region and beyond, potentially generating significant flows of asylum seekers affected by loss of their homes and livelihoods and prompting coastal States to review their migration and refugee laws. The trend towards greater numbers of asylum seekers transiting ocean areas in the South China Sea and South East Asia has a range of security implications for governments developing policy responses. There is accordingly likely to be greater public pressure on governments of transit and destination States such as Malaysia, Indonesia, Australia and New Zealand to tighten border security and devote more law enforcement and immigration resources to the detection and processing of boat people (Barnett and Adger 2007). The likely higher numbers of vulnerable displaced persons will provide additional opportunities for people smuggling networks to flourish, with commensurate pressures on police and prosecution authorities to target and deal with offenders. The hazardous nature of people smuggling operations, which entail movement by sea in overcrowded and unseaworthy vessels, are also likely to impose extra responsibilities on already overstretched regional navies and coastguards to respond to distress calls and escort people smuggling vessels to processing centres (Abbott 2008; CENTRA 2010).

The increased frequency and severity of weather-related disasters linked to climate change is likely to lead to loss of dwellings and the spread of disease in the Asia Pacific region and the need for augmented disaster relief services (Jasparro and Taylor 2008; CENTRA 2010). In the face of such severe weather events, the high volumes of shipping transiting the region may become more vulnerable to accidents at sea, and offshore energy installations may be more prone to damage and less able to provide reliable sources of energy (CENTRA 2010).

Fisheries scientists have predicted that the migration, depletion or even collapse of major fisheries as a result of rising ocean temperatures and increases in ocean acidity caused by global warming in addition to other factors such as overfishing will result in declining food security in regions that are heavily dependent on fish as a key source of protein for their populations (WorldFish Center 2006; FAO 2008; Allison et al. 2009). Indeed, the marine living resources of the Asia Pacific are critical to the food security of coastal populations numbering in the hundreds of millions (Schofield et al. 2011). Dwindling food and energy resources may provide a catalyst for inter- and intra-State disputes in the Asia Pacific region leading to a less predictable and secure maritime environment for trading and military access (Abbott 2008; Dupont 2008; CENTRA 2010). This book examines these diverse security effects, and the legal and policy initiatives being canvassed in the Asia Pacific region to tackle the maritime security implications of climate change.

5. CLIMATE CHANGE MITIGATION AND THE OCEANS

As well as the effects of climate change itself, marine geo-engineering schemes to mitigate the adverse impacts of climate change, such as ocean fertilization, pose significant risks to the marine environment that may lead to a less productive marine environment in the long-term (Shepherd 2009). In the process of adapting to the ravages of climate change, States and their nationals are already engaging in climate change mitigation activities with potentially adverse impacts on the marine environment. Similarly, maritime-oriented sectors such as the shipping industry are themselves turning their efforts towards minimizing the impact of their activities. International organizations are just beginning to grapple with the environmental regulation of activities, such as offshore carbon capture and ocean fertilization, and many coastal States have yet to incorporate appropriate precautionary measures and environmental impact assessment processes into their domestic legislation and policy. This book reviews progress at the global, regional and national levels in developing and implementing these measures.

6. ORGANIZATION AND CONTENT

As well as this introductory chapter, the book contains nine chapters examining the issues surrounding climate change and the oceans from several key perspectives. Chapters 2–4 explore the physical effects of climate change on the marine environment, fish stocks and marine biodiversity of the Asia Pacific region, and the legal and policy implications of these changes. Chapter 5 analyses the role of the Southern Ocean and the Antarctic region in the global climate and the broader legal and policy consequences of climate change impacts in these areas for other regions, including the Asia Pacific region. Chapter 6 analyses the complex issues surrounding maritime jurisdiction and climate change and the conundrums this poses for presently established maritime limits and boundaries, suggesting a number of response options open to coastal States in this context. Chapter 7 highlights the multidimensional security consequences of climate change impacts on the oceans, including increased people flows, threats to food security and loss of livelihood for many communities in the Asia Pacific region. Chapter 8 examines the ramifications of these broad security challenges for maritime security forces in the region and the potential transformations needed in force structures and postures to tackle these developments. Chapter 9 focuses on the topical issue of greenhouse gas (GHG) emissions from global shipping, and the regulatory framework being developed by the International Maritime Organization (IMO) to mitigate this significant contributor to global emissions. Finally, Chapter 10 describes the emerging array of climate change mitigation activities associated with the oceans, and global initiatives to regulate these activities. A brief synopsis of each of these chapters follows.

Chapter 2 by Ove Hoegh-Guldberg sets the scene for the chapters that follow. Hoegh-Guldberg explores the influences of global warming and ocean acidification on Pacific coastal and oceanic ecosystems, and the implications of these changes for the economy and well-being of people throughout the Asia Pacific region. The importance of coastal and ocean ecosystems in underpinning the food and livelihoods of Asia Pacific people due to the dominance of low-lying coastal areas and islands within the geography of the region is explained, as well as the highly productive nature of ecosystems such as coral reefs, mangroves and seagrass meadows. These ecosystem services include food, fisheries, building materials, medicine, tourist income and the indirect but important benefits derived from the stabilizing and protective function of ecosystems such as mangroves and coral reefs on coastal areas. The author also points to the benefits derived from resources within the Asia Pacific region associated with the open ocean such as tuna, which provides substantial revenue to

Asia Pacific States. Based on his discussion of future scenarios for the Asia Pacific region, Hoegh-Guldberg predicts that under the influence of global warming and ocean acidification, coral-dominated reef systems, mangroves and seagrass beds will contract by as much as 50 per cent, which will directly affect the temperature-dependent distribution of pelagic fish such as tuna. In addition there will be more frequent damaging storm systems and a rise in sea levels of around 10–15 cm. He concludes that marine ecosystems within the region will degrade further if the trends towards atmospheric CO_2 concentrations of 450 ppm and global temperatures of more than 2°C above the pre-industrial level are not halted, and recommends the adoption of a number of strategies to address these conditions in the Asia Pacific region. The first proposal is to reduce the impact of local stresses such as overfishing, coastal land degradation and marine pollution by increasing the resilience of coastal ecosystems to climate impacts such as storm damage, mass coral bleaching and heat stress. Secondly, in the management of fisheries, the suggestion is to create evolving, dynamic and 'adaptive' frameworks capable of responding to the changing conditions, challenges and opportunities arising from the shifting physical and chemical conditions throughout the Asia Pacific region. Thirdly, greater cooperation is advocated across nations in the management of pelagic fish stocks in response to changing conditions causing the movement of fish stocks. Finally, there needs to be greater cooperation across national borders, with the establishment of regional networks of expertise and capability to respond to the enormous challenges posed by climate change.

Richard Kenchington and Robin Warner focus on legal and policy responses to the impacts of climate change on marine biodiversity both globally and in the Asia Pacific region in Chapter 3. They point to some of the key effects of climate change on marine biodiversity, including increasing water temperature; changes in the chemical properties of seawater; sea-level rise; increased frequency, severity and range of severe weather events; and increasing thermal and other stresses on species due to all these factors. They discuss the difficulty in separating these effects from other anthropogenic impacts such as the overexploitation of fisheries and marine resources; coastal habitat destruction; and operational and catastrophic accidental pollution arising from marine industries, shipping, and land and freshwater uses. In terms of legal and policy responses they underscore that this amalgam of impacts on marine biodiversity, including climate change effects, requires multi-sectoral integration of policy and management within jurisdictions, coordination with adjacent and linked jurisdictions, and regional and international mechanisms to address areas beyond national jurisdiction. The chapter explores some of the key approaches to conserving and managing marine biodiversity in the face of threats such as climate change. These include marine protected area tools for biodiversity conservation and habitat protection, multi-use management tools that balance conservation and socioeconomic uses, sustainable use marine-resource management tools, primarily for extractive use and culture/ecological/social protection reserves, especially for indigenous and traditional non-indigenous communities. The chapter traces the development of legal and policy frameworks for the conservation of biodiversity and their linkages with climate change effects at the global, regional and national levels. The discussion concludes that ongoing efforts to address the impacts of climate change on marine biodiversity will benefit from the development of more integrated legal and policy frameworks for the conservation and management of biodiversity across national boundaries and beyond national jurisdictions.

In Chapter 4 Pia Winberg, Ana Rubio and Lisa Kirkendale focus on the vital need to sustain marine living resources for food production despite adverse climate change effects, examining subsistence, artisanal and global fisheries and aquaculture operations across different socioeconomic conditions in the Asia Pacific. They point out that collectively the Asia Pacific includes the three largest seafood-producing regions globally: the Northwest Pacific, the Southeast Pacific and the Western Central Pacific. Millions of people in both Asia and the Pacific rely on seafood production for subsistence as well as for local and export markets. They discuss the additional complexities involved in maintaining sustainable fisheries and aquaculture operations with the advent of climate change. Changes include altered marine food chains, shifting productivity, changes to the breeding patterns of commercial species and changes in their distribution – all leading to variable and erratic production. Three main categories of climate change effects – physical, chemical and ecological/biological – are substantiated and translated into a concrete suite of impacts on fisheries and aquaculture in the Asia Pacific region. The impacts predicted include changes in production and types of yield, growth and mortality effects, species shifts, increased pests and fouling, decreases in genetic diversity, lack of feed and reduced harvest and market regularity. A range of response options to each of these impacts is considered, and the policy implications of the options are synthesized at the end of the chapter.

Chapter 5 by Marcus Haward first addresses the critical role of the Southern Ocean and Antarctica in the global and regional climate. The winter–summer seasonal pattern of sea-ice formation and loss, which more than doubles the ice-covered area of Antarctica, is discussed as a driver in the global climate system with important links to the region's ecosystem. The role of the Southern Ocean as a major store or sink of the world's carbon and as an integral part of the carbon cycle where atmospheric CO_2 is absorbed by chemical and biological processes is also reviewed. The chapter then examines the regimes and instruments that govern the Southern Ocean, exploring the impact of climate change on the operation of these. Thus addressed are the Antarctic Treaty, the Protocol to the Antarctic Treaty on Environmental Protection and the Convention for the Conservation of Antarctic Marine Living Resources, as well as key global instruments such as the United Nations Convention on the Law of the Sea, the Convention on Biological Diversity, the United Nations Framework Convention on Climate Change (UNFCCC), and shipping instruments such as the International Convention on Prevention of Pollution from Ships (MARPOL 73/78) and the Convention and Protocol on the Prevention of Marine Pollution by Dumping of Wastes and Other Matters. Haward explains that the science-based focus of the Antarctic treaty system has been critical in supporting new research on the impacts of climate change on the Southern Ocean and the global oceans. The chapter concludes with observations on the influence of the intersection between science, law and policy in addressing climate change, drawing on the example of the Southern Ocean.

In Chapter 6 Clive Schofield and Andi Arsana explore one of the most notable impacts of climate change: sea-level rise. This phenomenon has led to concerns over the inundation of extensive low-lying coastal territories, with disastrous implications for coastal populations, infrastructure and environments. These concerns are underscored by longstanding trends towards the concentration of global population and activity in areas proximate to the coast such that an estimated 60 million people live within 1 m of present sea levels. Alarm has also been raised over the potential disappearance of entire islands and States entirely composed of islands, forcing island inhabitants to flee. Sea level rise also has significant potential to reduce the spatial extent of national claims to maritime jurisdiction—this maritime dimension of the impact of sea level rise provides the primary focus for discussion. The potential for losses to maritime claims because of sea-level rise emerges from the fundamental link between baselines – predominantly represented by low-water lines along the coast – and the limits of maritime jurisdiction zones. Further, the chapter highlights how sea-level rise is both spatially and temporally highly uneven, and notes that firm answers to the critical questions of how much and how swiftly global sea levels may rise presently remain uncertain – an issue that largely stems from the complexities inherent in measuring sea-level rise. The critical link between baselines and limits of maritime jurisdiction zones is then explored, with an emphasis on the potentially uneven ways sea-level rise could affect the extent of maritime claims. Options open to coastal States with a view to retaining their present maritime entitlements are then reviewed, with an emphasis on the various ways maritime limits can be defined and potentially stabilized.

Stuart Kaye considers in Chapter 7 how two key issues for the international community, maritime security and climate change, are related, and why the impacts of climate change may increasingly inform the nature of responses to maritime security. He posits that greater pressure on fisheries due to climate change may exacerbate a number of maritime security issues. These include potential increases in illegal fishing and bilateral disputes over marine living resources. Consequently, some States in the Asia Pacific region are more likely to adopt robust maritime law enforcement policies to protect precious and dwindling fish stocks in marine areas under their jurisdiction, as well as in those areas where they have potential claims to marine living resources. The converse position is that those States in the region with limited maritime law enforcement capacity to protect their fish stocks are likely to become more vulnerable to illegal fishing. Kaye also raises the issue of intensifying pressures on populations in the Asia Pacific region affected by sea-level rise to relocate. He foreshadows a more restrictive view being taken by developed States in the Asia Pacific region of their international law obligations to asylum seekers resorting to illegal migration by sea. In conclusion, Kaye predicts that the two issues of maritime security and climate change are likely to become more interlinked in the future in the Asia Pacific region and beyond.

In Chapter 8 Chris Rahman addresses the implications of climate change for maritime security forces including navies, coastguards and other maritime law enforcement agencies such as customs authorities. First, he considers the potential for climate change to create negative conditions for national or regional security. He highlights the negative political consequences that could arise as a result of the physical impacts of climate change, such as the loss of land for dwellings and pressure on natural resources. Although such conditions do not themselves necessarily lead to violent conflict requiring military intervention, Rahman suggests that, realistically, they could lead to problems of disorder on land and sea that require humanitarian assistance and law enforcement. In the Asia Pacific region Rahman suggests that this is likely to result in increased illicit activities at sea such as illegal fishing, smuggling, illegal migration, piracy and sea robbery. Secondly, he analyses the future demands that will be imposed on maritime security forces as a consequence of climate change and the possible implications for force structure development. Rahman focuses on the example of the US Navy as the organization that has contributed most to strategic thinking on climate change, and its implications for maritime strategy, policy and plans. The US Navy's Climate Change Roadmap identifies conditions of particular naval interest, including sea level rise, the frequency and intensity of tropical storms, rainfall patterns and concentrations of GHGs. These conditions in US Navy areas of influence could translate into greater demands for certain types of naval tasks such as maritime security, humanitarian assistance and disaster relief, defence support of civil authorities, maritime domain awareness, search and rescue, and strategic sealift. Rahman reviews the types of seaborne and airborne assets that will be needed to tackle these tasks. Thirdly, he discusses the internal challenges confronted by maritime security forces in reducing GHG emissions from their ships and other vehicles, and in attaining greater energy efficiency. Finally he suggests that the overall impact of climate change on maritime security forces is likely to be gradual rather than abrupt, leaving time for maritime security forces to invest in basic scientific research and marine environmental assessment, the development of new technological solutions to emissions reductions and better national and regional security risk assessments.

Chapter 9 by Stathis Palassis considers the broad problem of reducing GHG emissions from the burgeoning international shipping sector. The *IPCC Fourth Assessment Report: Climate Change 2007* estimated the contribution of international shipping to global GHG emissions in 2007 at 2.7 per cent. Palassis reviews the international legal framework for the reduction of GHG emissions, and the prevention, reduction and control of vessel source pollution, exploring how these two areas of international regulation have become interlinked. He discusses the first steps towards regulating atmospheric pollution from international shipping by the IMO through Annex VI to MARPOL 73/78, and the ambit of pollutants covered by these measures. He then traces the IMO's continuing efforts to reduce GHG emissions from international shipping through a combination of technical, operational and market-based measures. In particular, he canvasses the debates that have emerged in the IMO between developing and developed States over their relative responsibilities to reduce the GHG emissions of their shipping fleets. Palassis views the essence of the debate revolving around the role of developing States in reducing their GHG emissions as their strong reliance on applying the principle of common but differentiated responsibility and their questioning of the appropriateness of applying uniform international standards, which they see as economically disadvantageous. Developing States have also challenged the appropriateness of the IMO as a competent international organization and questioned whether amending Annex VI of MARPOL 73/78 to include GHG emissions is the correct regulatory strategy, identifying the UNFCCC and its COP mechanism as the more appropriate forum to discuss these issues. On the other side of the debate, Palassis explains that developed States view the IMO as the competent international organization within which to promulgate rules for the reduction of GHG emissions through the adoption of uniform standards under Annex VI of MARPOL 73/78. He tracks progress by the IMO in the introduction of energy efficiency measures for the international shipping sector, and discusses the complex legal and political challenges involved in introducing market-based measures designed to reduce GHG emissions from this growing transport sector.

Finally, in Chapter 10, Rosemary Rayfuse and Robin Warner examine the regulatory frameworks emerging for a range of CO_2 sequestration methods that either involve the oceans or affect the marine environment. They explain that considerable financial and technological investment has already been made in offshore carbon capture and storage (OCCS), which seeks to capture CO_2 from point sources of emissions and sequester it in sub-seabed geological formations. Other methods seek to boost the capacity of the oceans to capture and absorb atmospheric CO_2 either by depositing substances or wastes into the sea or depositing structures or devices into the ocean to increase the production of organic material in the surface ocean, thereby promoting increased draw down of photosynthesized carbon to the deep ocean. Both OCCS and marine geo-engineering schemes have the potential for yet unknown and possibly adverse effects on the marine environment. The chapter focuses on the legal challenges posed by these schemes, in particular ocean fertilization, and the progress that has been made in regulating these activities at global and national levels. It describes these processes, their objectives and their potential impacts on the marine environment. It discusses the applicability of existing international law principles to these processes and the regulatory gaps and ambiguities that exist. The emerging policy and regulatory framework for these processes at the global and national level in Australia is examined. Finally, Rayfuse and Warner conclude that the relatively advanced and permissive regulatory position developed for OCCS represents a significant contrast to the embryonic and decidedly cautious regulatory situation emerging for other marine geo-engineering methods designed to capture CO_2.

7. INNOVATIVE AND MULTILATERAL SOLUTIONS

Climate change poses a complex mix of threats to the environmental integrity and security of the marine and coastal areas of the Asia Pacific region, many of which are highlighted here. Combating these threats will demand innovative and multilateral solutions from regional States in cooperation with relevant regional and global organizations. Some of this cooperation is already occurring, but will require extension and ongoing development to reverse some of the adverse impacts of climate change on the biodiversity and security of the region. This book examines some of the collaborative mechanisms that could be implemented among States and regional organizations to alleviate some of the worst effects of climate change on marine biodiversity and maritime security in the Asia Pacific region, and considers progress towards achieving these mechanisms. These include joint development and conservation arrangements for fisheries and the introduction of transboundary marine environmental protection measures such as marine protected areas and transboundary environmental impact assessment of proposed activities and projects. Efforts are also under way within the region towards the settlement or at least management of contentious maritime disputes, as evidenced by the maritime joint zones and arrangements alluded to above. Multilateral mechanisms for hydrocarbon exploitation and ocean energy development could avert future energy supply downturns. The establishment of further cooperative maritime surveillance and enforcement arrangements among States in the Asia Pacific region may also assist in combating the increase in transnational criminal activity at sea. Enhanced criminal justice cooperation relationships among States in the region would complement these cooperative law enforcement efforts at sea. Cooperative resettlement arrangements between regional and extra-regional States for populations displaced by sea level rise or natural disaster will also be an important component of any plan to tackle the adverse security implications of climate change.

The negative impacts of climate change, combined with the other regional insecurity factors already outlined, are highly likely to provoke more competition for declining marine resources such as fish and untapped marine resources such as offshore hydrocarbons. There is already ample evidence of tension and active confrontation among some South East Asian States, for example over access to the hydrocarbon resources speculated to lie under the South China Sea. This 'oil factor' can also be seen as a driver in Indonesia and Malaysia's tensions on overlapping maritime claims – something not unconnected to control over the so-called 'Ambalat block', a potentially hydrocarbons-rich area of seabed in the Celebes Sea off eastern Borneo. These simmering disputes highlight the urgency of negotiating creative solutions to maritime disputes to enable equitable access to and distribution of available offshore resources. Many analysts have suggested that joint development arrangements be negotiated to facilitate exploitation of the hydrocarbon resources of the South China Sea, with the resolution of territorial sovereignty claims deferred to accommodate such exploitation. There are some models for this type of arrangement, such as the 2002 Timor Sea Treaty and the 2006 Certain Maritime Arrangements Treaty between Australia and Timor L'Este, although the number of claimants to the Spratly Islands amplifies the challenges involved in achieving an agreement acceptable to all parties. Consequently, as attractive as joint development for the South China Sea appears on paper, unfortunately this does not appear to be a likely outcome in the near term, though this does not eliminate the possibility of gradual, incremental progress to resolve or at least manage disputes (Schofield et al. 2011).

In the fisheries sphere such positive developments are illustrated by the fact that there are already some joint conservation and management arrangements in the South China Sea and South East Asia, and these practical joint endeavours may assist in preventing potential disputes among regional States over declining fish stocks and in adapting to the changes in fish stocks that could occur due to climate change. For instance, China and Vietnam negotiated two agreements to conserve, manage and exploit the fisheries resources of Beibu Gulf: the 2000 Agreement between the Government of the People's Republic of China and the Socialist Republic of Vietnam on Fishery Cooperation in Beibu Gulf and a 2004 protocol to the earlier agreement. The agreements are administered by a Joint Fishery Committee with members from both parties. Three zones are designated: the Common Fisheries Zone, the waters in transitional arrangements, and the buffer zone for small sized fishing boats. The Common Fisheries Zone covers 33 500 km^2 – 27.9 per cent of the Gulf's total area. The parties are committed to undertaking long-term fisheries cooperation in this area over the 12-year term of the agreement, with a further three-year extension envisaged in Article 22 of the 2000 agreement, which involves joint development of conservation and management measures to conserve, manage and sustainably exploit the living resources of the zone. The number of vessels with access to the zone is determined annually, taking into account allowable catches established on the basis of joint regular surveys of fishery resources, the impact on respective fishing activities of both parties, and the need for sustainable development of the resources while respecting the principle of equality and mutual benefit. The waters in transitional arrangements zone was established as an interim measure to limit the number of fishing vessels of each party that could access certain areas of their respective exclusive economic zones for four years after the 2000 agreement entered into force. The buffer zone for small sized fishing boats (no longer than 15 m or with engines of no more than 60 horsepower) was established to avoid disputes caused by the inadvertent illegal entry of small sized fishing boats of one party into the territorial sea of another. Under the agreement if either party finds small sized fishing boats of the other party in its buffer zone waters, it may warn the boats to leave or take necessary measures to repel them from the area without using force or detention and arrest. Cooperative arrangements of this type, including those mentioned in the Gulf of Thailand with respect to seabed oil and gas and in the East China Sea in relation to fisheries, are potentially critical lynchpins in mitigating the extra burden created by climate change effects among States in the Asia Pacific region.

8. CONCLUSIONS

While climate change may not be the only catalyst for marine biodiversity loss and regional insecurity in the Asia Pacific region, it has a compounding effect on other causes of these problems. Climate change represents a potent exacerbating factor of other biodiversity loss and regional insecurity drivers such as overfishing, destructive fishing practices, poverty, fragile social circumstances and existing transnational crime networks. Similarly, the impacts of climate change on maritime jurisdictional issues threaten to complicate multiple already highly contentious and involved regional maritime disputes featuring broad areas of overlapping and conflicting maritime claims. If climate change adaptation measures are not rigorously implemented the adverse effects of warming temperatures and ocean acidity on fish stocks, combined with overfishing and destructive fishing practices, could lead to the collapse of some fish stocks in the region, precipitating a food security crisis. Reinforcing the regional measures already in place to combat transnational crimes such as people smuggling and illegal fishing will be important in reducing the vulnerability of coastal populations displaced and marginalized by sea-level rise and severe weather events associated with climate change. Re-examining the maritime boundary and resource-related disputes that exist in the Asia Pacific region to establish whether creative solutions can be found to maximize the benefits for all interested stakeholders can only assist the region in addressing the resource shortages that climate change and other factors will inevitably bring. Some coping mechanisms and multilateral initiatives are already in place to address these developments. However the adverse portents of climate change documented by the IPCC and other international bodies suggest that collaborative initiatives to mitigate its negative effects on marine biodiversity and food and human security need to be strengthened at all levels.

REFERENCES

Abbott, Chris (2008), *An Uncertain Future: Law Enforcement, National Security and Climate Change*, Madrid, Spain: Fundacion para las Relaciones Internacionales y el Dialogo Exterior, accessed 22 April 2012 at: http://www.fride.org/download/COM_Law_Security_Climate_EN_feb08.pdf.

Allison, Edward H., Allison L. Perry, Marie-Caroline Badjeck, W. Neil Adger, Katrina Brown, Declan Conway, Ashley S. Halls, Graham M. Pilling, John D. Reynolds, Neil L. Andrew and Nicholas K. Dulvy (2009), 'Vulnerability of national economies to the impacts of climate change on fisheries', *Fish and Fisheries* **10** (2), 173–196.

Barnett, Jon and W. Neil Adger (2007), 'Climate change, human security and violent conflict', *Political Geography*, **26** (6), 643–655.

Bradford, John F. (2005), 'The growing prospects for maritime security cooperation in Southeast Asia', *Naval War College Review*, **58** (3), 63–86.

Caron, David D. (1990), 'When law makes climate change worse: rethinking the law of baselines in light of a rising sea level', *Ecology Law Quarterly*, **17** (1990), 621–653.

CENTRA Technology, Inc. and Scitor Corporation (2010), *Southeast Asia: the Impact of Climate Change to 2030: Geopolitical Implications*, Conference Report CR 2010-02, January, US National Intelligence Council, accessed 22 April 2012 at: http://www.dni.gov/nic/PDF_GIF_otherprod/climate_change/cr201002_southeast_asia_pacific_islands_climate_change.pdf.

Doyle, Thomas W., Richard H. Day and Thomas C. Michot (2010), *Development of Sea Level Rise Scenarios for Climate Change Assessments of the Mekong Delta, Vietnam*, Reston, Virginia: U.S. Geological Survey Open-File Report 2010–1165.

Dupont, Alan (2008), 'The strategic implications of climate change', *Survival: Global Politics and Strategy*, **50** (3), 29–54.

Food and Agriculture Organization (FAO) (2008), *Report of the FAO Expert Workshop on Climate Change Implications for Fisheries and Aquaculture*, FAO Fisheries Report No. 870, April, Rome, Italy: FAO, accessed 22 April 2012 at: ftp://ftp.fao.org/docrep/fao/010/i0203e/i0203e00.pdf.

Freestone, David (1990), 'International law and sea level rise', in Robin Churchill and David Freestone (eds), *International Law and Global Climate Change*, London, UK and Dordrecht, The Netherlands: Graham and Trotman/Martinus Nijhoff.

IPCC (2007), *Contribution of Working Group II to the Fourth Assessment Report of the Intergovernmental Panel on Climate Change, 2007*, New York, NY, USA: Cambridge University Press, accessed 22 April 2012 at: http://www.ipcc.ch/publications_and_data/publications_ipcc_fourth_assessment_report_wg2_report_impacts_adaptation_and_vulnerability.htm.

Jasparro, Christopher and Jonathan Taylor (2008), 'Climate change and regional vulnerability to transnational security threats in Southeast Asia', *Geopolitics*, **13** (2), 232–256.

Mekong River Commission (MRC) (2010), *Reducing the Impact of Climate Change on the Mekong Basin: the MRC's Climate Change and Adaptation Initiative*, MRC Technical Paper No. 34, December, Phnom Penh, Cambodia: MRC, accessed 22 April 2012 at: http://www.mrcmekong.org/about-the-mrc/programmes/climate-change-and-adaptation-initiative/.

Paskal, Cleo (2007), *How Climate Change is Pushing the Boundaries of Security and Foreign Policy*, Chatham House Briefing Paper, June, London, UK: Chatham House, accessed 22 April 2012 at: http://www.chathamhouse.org/sites/default/files/public/Research/Energy,%20Environment%20and%20Development/bp0607climatecp.pdf.

Prescott, Victor and Clive H. Schofield (2005), *The Maritime Political Boundaries of the World*, Leiden, The Netherlands and Boston, MA, USA: Martinus Nijhoff Publishers.

Schofield, Clive H. (2009), 'Shifting limits? Sea level rise and options to secure maritime jurisdictional claims', *Carbon and Climate Law Review*, **3** (4), 405–416.

Schofield, Clive H., Ian Townsend-Gault, Hasjim Djalal, Ian Storey, Meredith Miller and Tim Cook (2011) *From Disputed Waters to Seas of Opportunity: Overcoming Barriers to Maritime Cooperation in East and Southeast Asia*, National Bureau of Asian Research Special Report No.30, July 2011.

Shepherd, John et al. (2009) *Geo-engineering the Climate: Science, Governance and Uncertainty*, London: Royal Society.

Smith, Paul J. (2007), 'Climate change, mass migration and the military response', *Orbis*, **51** (4), 618–619.

Willis, B.L., S. Dove, O. Hoegh-Guldberg, J.M. Lough, M. McCulloch, P.L. Munday, M.S. Pratchett, R. Salm and M.J.H. van Oppen (2008), 'Climate change threats to the Coral Triangle', Draft Background Paper, Version 2, Coral Triangle Initiative Townsville Workshop November, accessed 22 April 2012: http://www.coralcoe.org.au/events/ctiworkshop/papers/8%20Climate%20 Change.doc.

WorldFish Center (2006), *The Threat to Fisheries and Aquaculture from Climate Change*, Policy Brief, Penang, Malaysia: WorldFish Center, accessed 22 April 2012 at: www.worldfishcenter.org/resource_centre/ClimateChange2.pdf.

2. Implications of climate change for Asian-Pacific coastal and oceanic environments

Ove Hoegh-Guldberg

1. INTRODUCTION

People in the Asia-Pacific region live in an intimate proximity to coastal and oceanic environments. This relationship has dictated a high degree of dependence on the ecosystem goods and services provided by coastal and oceanic ecosystems. Whether it is coral reefs, mangroves or pelagic fish stocks, Asian-Pacific people draw significant resources in terms of food, livelihoods, cultural value and coastal protection from the marine resources that surround them (Moberg and Rönnbäck 2003; Ronnback 1999). The strong linkages between local climates, coastal ecosystems and human communities underscore the importance of maintaining these relationships while moving towards a sustainable future (Woodward et al. 1998).

Broadly speaking, the marine environment is separated into two distinct although linked environments. The first represents coastal environments common to both large and small islands, while the second is composed of the open waters of the Pacific Ocean. The two environments are linked by physical, chemical, biological and economic events and processes. Understanding the nature of these linkages forms an important part of the broader understanding of how Asian-Pacific coastal and oceanic environments are likely to change under the influence of both local (e.g., fishing, unsustainable coastal zone management) and global (e.g., ocean acidification and global warming) pressures (Woodward et al. 1998). In this chapter, the implications of current and future changes to Pacific coastal and oceanic ecosystems is explored, with particular attention to how these changes are likely to influence the economy and well-being of people throughout the Asia-Pacific region.

2. CHARACTERISTICS OF ASIAN-PACIFIC COASTAL AND OCEAN ECOSYSTEMS

The Asia-Pacific region is dominated by the Pacific Ocean, which is the largest ocean on our planet. This vast ocean covers 165.25 million km^2 which is approximately one-third of the Earth's total surface. Within this region, the Pacific lines the coasts of four major continents (Australia, Eurasia, North and South America) and interacts closely with the waters surrounding the fifth continent, Antarctica. Given the geographical spread, the region also includes a multitude of cultures that are derived from European, Asian, Melanesian, Micronesian and Polynesian traditions. Many of these traditions, spread across hundreds of thousands of kilometres of low-lying coastline and encompassing over 25 000 islands, are strongly associated with the ocean.

Two major oceanic gyres dominate the Asia-Pacific region (see Figure 2.1). The clockwise North Pacific Gyre is driven westward by the trade winds, turning north near the Philippines to become the Kuroshio current and flowing towards the Arctic Circle before pushing down the West Coast of the United States. In the southern hemisphere, the south equatorial current flows from South America towards the southern coast of Papua New Guinea before moving southward down the east coast of Australia towards Antarctica and eventually flowing eastward towards South America once again. These current systems play major roles in determining the local conditions and weather experienced by the countries throughout the Asia-Pacific region. In this regard, powerful climate systems like the El Niño-Southern Oscillation (ENSO) ultimately drive periods of inundation or drought in countries like Australia, North Asia and the Americas. Variability in ENSO joins phenomena such as the monsoonal weather system in driving rainfall, productivity and ultimately the well-being of people within the Asia-Pacific region. These weather systems are determined by temperature differentials between land and sea and, consequently, the ocean temperature in the Indian and Pacific plays a particularly important role in determining the distribution and strength of the influence of ENSO and the monsoonal weather systems (Meehl et al. 2000).

Temperature also plays an important role in driving the productivity and biodiversity of ecosystems across the Asia-Pacific region, with ocean temperature being particularly important in defining the types of ecosystems found at particular locations within the Pacific. In this regard, the surface waters of the Pacific Ocean experience a wide range of ocean temperatures (0–33°C), playing host to a broad range of different marine ecosystems, from cold-water benthic systems dominated by kelp forests

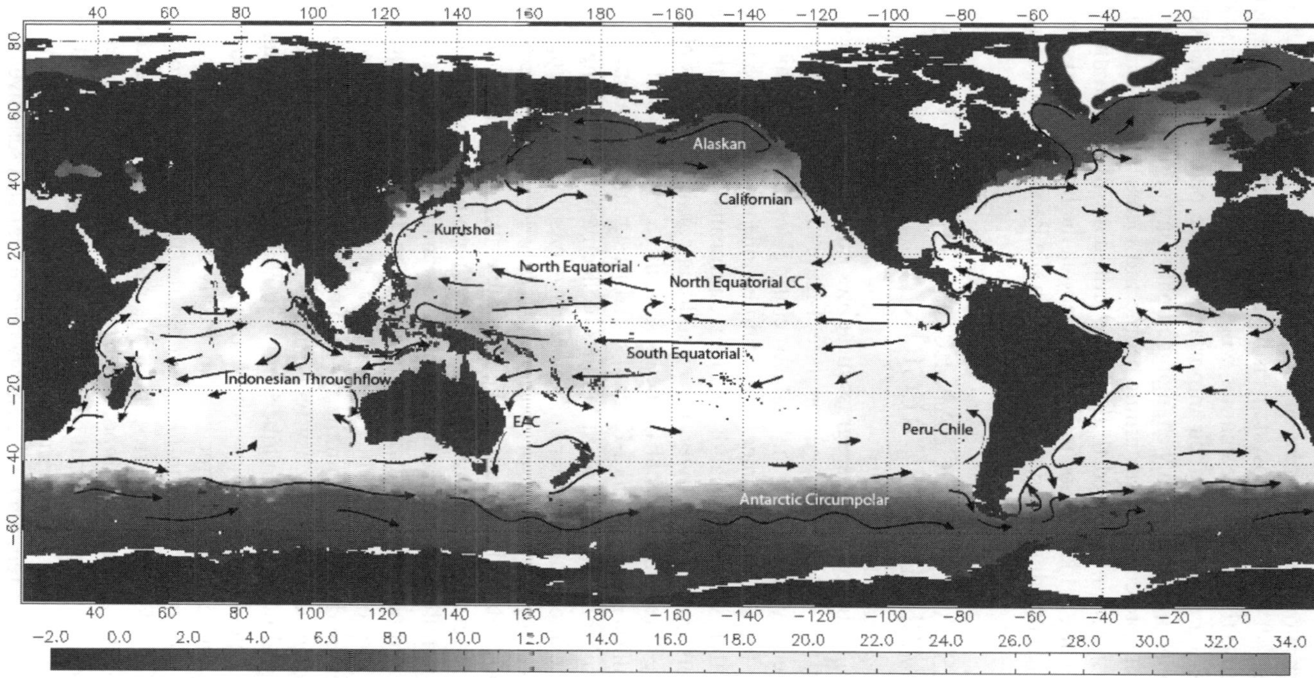

Figure 2.1 *Global sea surface temperature on 17 February 2011 (NOAA, NESDIS) and major ocean currents of the Asia-Pacific region*

at high latitudes, to the coral reefs, mangroves and seagrass meadows that flourish along coastal areas at lower latitudes. A full discussion of the complete set of Pacific ecosystems is beyond the scope of this chapter. Consequently, the rest of the chapter will be devoted to a discussion of tropical coastal ecosystems, their importance to the people of the Asia-Pacific region, and how rapidly changing climate may influence these ecosystems' ability to continue sustaining people and societies into the future. This discussion, while limited to the lower latitudes within Asia and the Western Pacific, is likely to be informative about the linkages and consequences of the rapidly changing climate within the Pacific region.

3. COASTAL AND OCEAN ECOSYSTEMS WITHIN THE ASIA-PACIFIC REGION

Coral reefs, mangroves and seagrass meadows dominate coastal regions throughout the Asia-Pacific region (see Figure 2.2). Each ecosystem is dominated by particular 'habitat structuring' organisms, which are responsible for the establishment and maintenance of three-dimensional habitats within which thousands of other organisms reside. In the case of coral reefs, calcifying activity of reef-building corals is responsible for forming the three-dimensional framework of an ecosystem that houses over one million species (Reaka-Kudla 1997). In a similar way, amphibious trees from a broad range of plant families (collectively called mangroves) structure the intertidal boundary of low-energy coastlines, providing food and resources for a multitude of additional species. Together, these organisms are particularly important as the ultimate drivers of tropical coastal marine biodiversity, which underpins the livelihoods and well-being of hundreds of millions of people within the Asia-Pacific region.

Coastal and ocean ecosystems are important in underpinning the food and livelihoods of Asian-Pacific people due to the dominance of low-lying coastal areas and islands within the geography of the region, as well as the highly productive nature of ecosystems such as coral reefs, mangroves and seagrass meadows (Hoegh-Guldberg et al. 2009b). In populous countries such as Indonesia, well over two-thirds of the population live within the coastal region and depend directly or indirectly on the goods and services of coastal ecosystems (Hoegh-Guldberg et al. 2009b). These ecosystem products include food, fisheries, building materials, medicine, tourist income and the indirect but important benefits derived from the stabilizing and protective function that ecosystems such as mangroves and coral reefs have on coastal areas. For example, mangroves and coral reefs play important roles in reducing wave energy before it affects sensitive coastal

Figure 2.2 *Enormous benefits are generated for people in the Asia-Pacific region as a result of coastal ecosystems. Examples are shown of (A) coral reefs, (B) mangroves, (C) seagrass meadows, and (D) pelagic ecosystems (photos O. Hoegh-Guldberg)*

areas; the benefits are substantial in terms of protecting human communities and their coastal infrastructure (Costanza et al. 1997; Moberg and Folke 1999; Moberg and Rönnbäck 2003; Orth et al. 2006).

One of the most important observations to make upfront about coastal ecosystems is that the area of these resources often greatly exceeds the land area in many Asia-Pacific States. The significance of these resources within the region has led to many Asia-Pacific leaders initiating programmes that focus on how to preserve these valuable ecosystems. One of the most recent was initiated by President Susilo Bambang Yudhoyono of Indonesia and the leaders of five other Asia-Pacific States. Together, these leaders articulated a vision which has become known as the Coral Triangle Initiative (Jompa 2008). The vision behind this initiative recognizes the importance of coastal ecosystems to the wealth and well-being of people from Indonesia, Malaysia, Philippines, Solomon Islands, Papua

New Guinea and Timor Leste, and articulates a strategy within which the multitude of local and global threats can be addressed (Hoegh-Guldberg et al. 2009a).

In addition to the goods and services provided by coastal ecosystems, additional benefits are derived within the Asia-Pacific region from resources associated with the open ocean. For example, fisheries associated with pelagic species such as tuna provide substantial revenue to Asia-Pacific States (Chand et al. 2004; Kompas and Che 2006). These valuable species are sensitive to changing environmental conditions, which can affect early life stages (often spent in coastal waters), and the distribution and abundance of adults (Kitagawa et al. 2000; Lehodey et al. 1997, 2003). As will be developed in subsequent sections of this chapter, future changes in the distribution and abundance of these species and associated fisheries may present significant challenges for some countries into the future.

4. LOCAL-SCALE THREATS TO THE SUSTAINABILITY OF COASTAL ECOSYSTEMS

Despite their importance to the people and economies throughout the Asia-Pacific region, coastal ecosystems are under serious pressure from a large range of human activities (Caldwell et al. 2009; Halpern et al. 2008). These problems stem from the large numbers of people living within coastal areas of the Asia-Pacific and their close dependence on marine resources (Burke et al. 2002; Spalding et al. 2001, 2003, 2007). The effect of people and their activities has led to declining water quality within coastal areas due to unsustainable land use within river catchments, physical damage through deforestation of mangroves and destructive fishing practices, and broad ecological changes resulting from the overharvesting of ecologically important species. Halpern et al. (2008) recently provided an integrated global map of the human impact on marine ecosystems, which revealed that the Asia-Pacific region includes areas that are among the most, and the least, affected by human activities (see Figure 2.3). Other broad surveys have concluded that coastal ecosystems such as coral reefs, mangroves and seagrass meadows are contracting at rates approaching 1–2 per cent per year (Bruno and Selig 2007; Duke et al. 2007; Waycott et al. 2009) as a result of local human activities.

When considering the effects of climate change on marine resources, it is often difficult to distinguish the influence of local factors from the impact of global climate change. In fact, it is important to emphasize the synergies between local and global factors. For example, reduced numbers of particular types of fish species (such as herbivores) can reduce coral reefs'

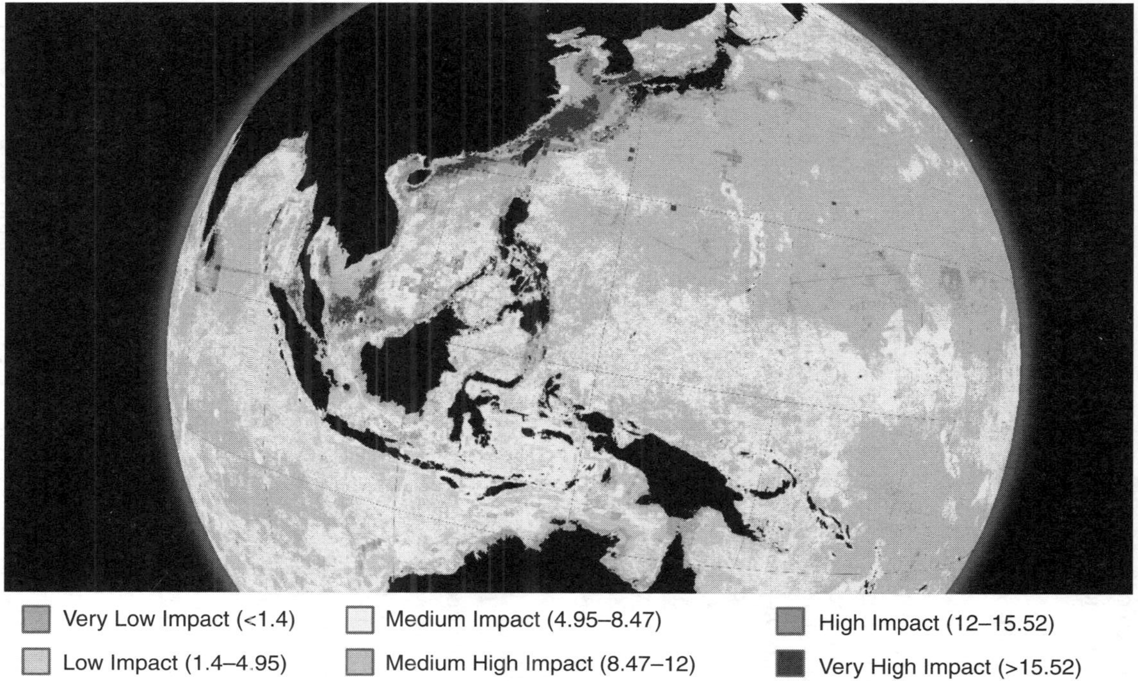

Source: After Halpern et al. 2008; adapted from Google Earth using 'Marine impacts' kml file (http://ebm.nceas.ucsb.edu/GlobalMarine/kml/marine_model.kml)

Figure 2.3 Human impacts on marine resources as a result of local and global stresses within the Asia-Pacific region

ability to recover from events in which thermal stress removes corals from the substrate (Hughes et al. 2003a, 2003b, 2007). These types of linkages are important to identify given that they may represent opportunities for improving the ecological resilience of coral reefs to growing stresses from rapid anthropogenic climate change.

5. CLIMATE CHANGE IN THE ASIA-PACIFIC REGION

The species and ecological processes that make up the coastal and oceanic ecosystems within the Asia-Pacific region are strongly influenced by both marine and surface climates (Griffiths et al. 2005; Meehl et al. 2000). Environmental changes arising from a shifting climate have the potential to influence a broad range of ecological variables, including the distribution and abundance of early life-history stages, and the growth of adult-phase organisms. Understanding how climate change is likely to affect species and ecosystem processes necessarily begins with an attempt to understand how conditions are changing and are likely to change. This section provides a broad summary of the principal environmental changes expected within the Asia-Pacific region because of rising fossil fuel emissions and other impacts on the global climate.

The evidence for the warming of the Earth's climate system is unequivocal (IPCC 2007b). The year 2010 was the hottest year on record, and each of the past three decades has been hotter than the one before (Hansen et al. 2010). The average global temperature has increased by 0.8°C since the advent of the Industrial Revolution (IPCC 2007b), which resulted in large amounts of fossil fuel carbon being ejected into the atmosphere. Long-term records that extend thousands of years into the past show a tight correlation between the recent rise of greenhouse gases such as fossil fuel derived carbon dioxide and average global temperature. Together with the large-scale alteration of forests and natural landscapes, these changes have changed the composition of the Earth's atmosphere and have enhanced the greenhouse effect. Global temperature has been increasing at a rate of 0.2°C per decade over the past 30 years (Hansen et al. 2010), resulting in changes to weather patterns and other fundamental parts of the land-sea-atmosphere system that have been relatively constant by comparison for thousands of years.

The oceans play a prominent role in the processes and patterns resulting from climate change. Over 90 per cent of the additional energy absorbed by Earth as a result of the inherent greenhouse effect has been absorbed by the upper 700 m of the world's oceans, resulting in a 14×10^{22} Joule increase in the amount of energy in the upper layers of the ocean since 1975 (Levitus et al. 2009). This is the equivalent amount of heat energy to that of detonating two atomic bombs per second. The addition of this extra heat has resulted in an increase in the temperature of the upper layers of the world's oceans of 0.6°C over the past 30 years. At the same time, approximately 30 per cent of the extra carbon dioxide added to the atmosphere has been absorbed by the ocean (Doney et al. 2009). Carbon dioxide reacts with water to create a dilute acid (carbonic acid), resulting in a shift in the abundance of important chemical species, such as carbonate ions (see Figure 2.4). These chemical changes have been collectively termed 'ocean acidification' due to the associated decrease in the pH value of ocean waters. The downward movement of ocean pH has been confirmed by a wide number of oceanographic surveys, with an average shift from pre-industrial values of 0.1 pH units and a current rate of change of 0.02 pH units per decade (Doney et al. 2009). These changes are occurring at faster rates at high latitudes due to the greater solubility of carbon dioxide in cool waters (Doney et al. 2009).

The additional energy that has been absorbed by the ocean over the past 100 years has resulted in a number of fundamental changes to the physical properties of the world's oceans. Sea ice that may have been present for over 700 000 years in the Arctic summer is rapidly disappearing (Wang and Overland 2009; Wang et al. 2002). In addition to driving the loss of summer sea ice, heating the upper layers of the ocean has resulted in thermal expansion of the ocean, which, in combination with the added melt water from glaciers and landlocked ice sheets, has increased sea level by approximately 20 cm since the nineteenth century. Data from satellite altimeters and other sources reveal an average global sea level rise of 3.3 ± 0.4 mm yr^{-1} (over the period 1993–2006), with a significant acceleration of sea level rise of 0.013 ± 0.006 mm yr^{-2} (Church and White 2006). Adding energy to the ocean also has implications for the intensity of storm systems. Warm oceans drive more intense storm systems (Stowasser et al. 2010) that, when combined with high seas, have serious implications for natural ecosystems and coastal human settlements. The recent increase in the frequency of destructive storm systems across the Asia-Pacific region is testimony to the relatively simple physical link between warming oceans, intensified hydrological cycles and intense storm systems (Knutson et al. 1998, 2010).

The question of how conditions will change within the Asia-Pacific region under rapid anthropogenic climate change is necessarily complex and requires an in-depth understanding of weather and oceanographic patterns which are relatively poorly understood. Long-term and high-resolution records are incomplete, and may not be available at all for some

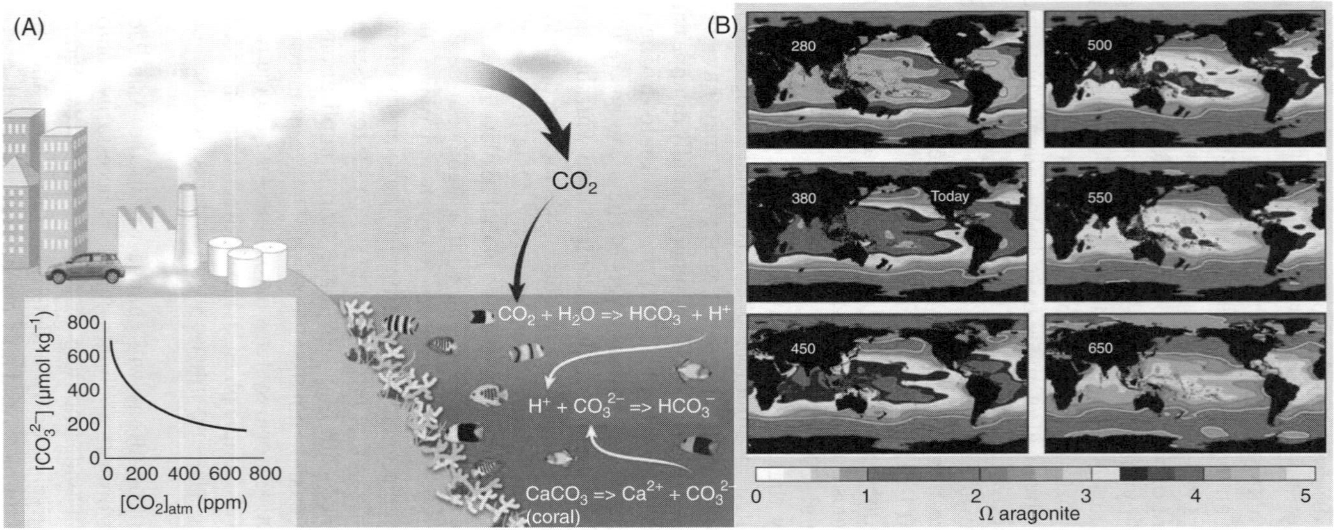

Source: Reprinted with permission of *Science Magazine*: Hoegh-Guldberg et al. 2007.

Figure 2.4 *Ocean acidification arises from the increased entry of carbon dioxide into the ocean as atmospheric levels of this greenhouse gas increases in the atmosphere. A. Once in the ocean, carbon dioxide reacts with water to create a dilute acid, which reacts with carbonate to produce bicarbonate. The reduced pH and availability of carbonate ions affects the ability of some corals and other marine calcifiers to form their calcium carbonate skeletons. B. Area suitable for carbonate coral reefs as a function of atmospheric carbon dioxide values higher than 500 ppm*

countries and sectors. However, there are several clear indicators, such as warming, acidifying and expanding oceans, that provide some insight into how the challenges associated with anthropogenic climate change within the Asia-Pacific region may unfold. Evidence that climate change is occurring in the Asia-Pacific is irrefutable. Sea temperatures have been increasing at rates of 0.1–0.4°C per decade, although with significant geographical variability (Peñaflor et al. 2009). At the same time, oceans have acidified by 0.1 pH units, and have seen a decline in the concentration of the key chemical ion species, carbonate, of around 30 per cent since the pre-industrial period.

Associated with these changes has been a series of rainfall anomalies throughout the Asia-Pacific region, with prolonged drought in some areas and record rain inundation in others (Hoegh-Guldberg et al. 2009b; IPCC 2007a). Changes to rainfall intensity may have significant ramifications for the coastal ecosystems, given the potential for these events to destabilize and move large amounts of sediment and nutrients off the land and into coastal environments. Much of the temporal and spatial resolution of these events is determined by the strength and phase of the ENSO, which has a profound influence on patterns of rainfall, drought, storm activity and other weather patterns across the Asia-Pacific region (Chou et al. 2010; Meehl et al. 2000). The recent record rainfall throughout the region in the first part of 2011, and again in 2012, underscores the important interactions between anthropogenic climate change and long-term patterns of variability. Understanding these changes will be critical for any successful attempt to anticipate and adapt to the consequences of climate change.

6. IMPACTS OF CLIMATE CHANGE ON MARINE ECOSYSTEMS

Biological systems are closely connected to the environment that surrounds them. The marine ecosystems that dominate the inshore waters of the Asia-Pacific are already showing responses to the changing physical and chemical conditions within the ocean. Habitat forming organisms such as reef-building corals, mangroves and seagrasses appear to be particularly vulnerable to changing ocean temperature, pH, sea level and other factors. While marine organisms and ecosystems may respond gradually to changes in the environment, there is considerable evidence that changes can be abrupt and non-linear. Understanding the changes occurring currently provides important insights into likely future changes in the marine ecosystems within the Asia-Pacific region. This information is important if we are to anticipate and respond effectively to the challenges that are likely to arise over the coming decades and century. The following discussion will explore the effects of climate change on four key marine resources, with a discussion of how changes in future conditions will affect the ability of these resources to support the ecosystem services that are so valuable to the people of the Asia-Pacific region.

6.1 Coral Reef Ecosystems

Coral reef ecosystems are prominent components of near-shore habitats across the tropical Asia-Pacific (Burke et al. 2002; Spalding et al. 2001). Coral reefs form complex ecosystems that provide ecological services such as food, income and coastal protection, and play a significant cultural role within most Asia-Pacific nations. Reef-building corals lie at the core of coral reef ecosystems. The carbonate framework they provide is contributed to by a range of other organisms (e.g., foraminifera and crustose coralline algae) to build the reef infrastructure that will provide the habitat for an estimated one to nine million species (Reaka-Kudla 1997). The resulting three-dimensional structure not only provides habitat, but also modifies coastal wave environments. Over very long periods, the activities of corals and other marine calcifiers lead to the accumulation of structures that may eventually form islands and barrier reef systems (Andréfouët et al. 2009).

Coral reefs thrive in the Asia-Pacific region between 25°N and 25°S in warm, relatively alkaline and sunlit locations. If temperatures drop below 18°C in the winter, carbonate coral reefs will fail to form. Further, coral reefs require values of aragonite saturation (a measure of carbon chemistry) that do not decrease below 3.2. Coral reefs are absent from coastal areas where light levels are low or where coastal waters are turbid (Spalding et al. 2001). While corals may persist in these regions, their ability to build and maintain carbonate reef structures is much reduced. While there may be corals within these regions, they do not establish the carbonate coral reef ecosystems that typify many coastal areas of the Asia-Pacific region (Andréfouët et al. 2009). However, once established, carbonate coral reef ecosystems play a critically important role in providing populations of fish and other life forms for food and income generating activities. In the tropical Pacific alone, local fisheries provide an estimated 50 to 90 per cent of the animal protein of coastal communities (Bell et al. 2009). Marine-based protein sources are similarly important within Southeast Asian countries such as Indonesia and the Philippines (Hoegh-Guldberg et al. 2009b).

Four major types of fisheries are associated with coral reef ecosystems (Pratchett et al. 2011):

- *Shallow sub-tidal and intertidal invertebrate fisheries*, which involve the harvest of a wide range of molluscs (e.g., giant clams, gastropods and octopus) and echinoderms (e.g., sea cucumbers and urchins). The harvest is performed primarily by women and children and it is thought to be an important source of household food, although this activity has gone largely undocumented, due to its casual, non-market nature.
- *Demersal fin fisheries*, which are comprised of reef fish caught using gillnets, hand lines or spears. The products of this fishery are used mainly as food for local people or may be sold at local markets to supplement income. This type of fishery is linked to the destructive trade in live fish for the Southeast Asian restaurant market or the tropical marine aquarium market, both of which have had significant impacts on coral reefs ecosystems across the Asia-Pacific region.
- *Targeted invertebrate fisheries*, which focus on non-fish species such as sea cucumbers, trochus, giant clams, spiny lobsters and other invertebrate species. These fisheries have provided important income for communities throughout the Asia-Pacific for at least 100 years. These fisheries provide important income for coastal communities, and may involve long trade routes throughout the region.
- *Near-shore pelagic fisheries*, which harvest stocks of Skipjack, Yellow-fin and Big-eye tuna via offshore industrial fishing fleets that may or may not be owned by the country being harvested. These fisheries also capture a large range of other species including Mahi-mahi, Oahu and Spanish mackerel, as well as a range of small pelagic fish species including mackerel, pilchards, anchovies, scads and flying fish. More recently, these fisheries have concentrated their efforts using Fish Aggregating Devices. Income earned from near-shore pelagic fisheries provides an important supplement to coastal food resources, and provides significant income earning opportunities for coastal communities throughout the Asia-Pacific.

As with the pelagic fisheries, coastal fisheries are fundamentally dependent on the health of coral reef ecosystems, which, along with associated mangroves and seagrass meadows, provide the fundamental habitats required by these fish and invertebrates species. Populations of these important species decline rapidly following the loss of reef-building corals and reef structure because of local stressors such as over-exploitation, destructive fishing and declining water quality (Burke et al. 2002).

Coral reefs provide a range of other services in addition to supporting fisheries. The tourist industry that based itself on the pristine values of coral reef ecosystems, while small in terms of total revenue generated, may provide significant income for many Asia-Pacific countries. Coral reefs, by providing the primary structure in which the famous Asia-Pacific 'mega-diversity' resides, play a critical role in supporting ecosystem-based tourism in the region. Further, much of the carbonate sand, highly valued by tourists, is generated by healthy coral reef ecosystems. Coastal protection provided by coral reefs ecosystems can also play a fundamentally important role in protecting human communities and infrastructure along the low-lying island coastlines of the Asia-Pacific. Estimating the value to coastal communities of reef ramparts where the energy of ocean waves is dissipated is difficult, but there is little doubt regarding the significance of this contribution. Further value can be gained from coral reef ecosystems through the exploitation of the mega-diversity of the Asia-Pacific region for new biochemicals and medical products. As mentioned, healthy coral reef ecosystems provide habitat to more than 1 million species, providing a rich field for this kind of exploitation.

6.1.1 Mass coral bleaching and mortality

Over the past three decades, it has also become increasingly clear that coral reefs are particularly sensitive to global stresses such as rising sea temperatures and acidifying oceans (Hoegh-Guldberg et al. 2007). Understanding the present-day responses of reef-building corals to these changes provides insight into how coral reef ecosystems are likely to change in the future. Central to the sensitivity of reef-building corals to the chemical and physical changes around them is the mutualistic symbiosis that corals form with small plant-like organisms known as dinoflagellates (genus *Symbiodinium*). *Symbiodinium* impart the brown colour to the tissues of corals and are able to photosynthesize while located within the endodermal cells of the host coral. Photosynthetic products generated by *Symbiodinium* are passed to the coral host, which provides *Symbiodinium* with access to inorganic nutrients (from host waste metabolism) that are otherwise rare in tropical waters. The energy transferred by *Symbiodinium* is used for the growth, reproduction and calcification of the coral host (Muscatine 1990). Due to this abundant energy, reef-building corals are able to grow rapidly and produce an expansive skeleton of calcium carbonate. Indeed, this symbiotic relationship is the key to the success of the diverse and highly productive coral reef ecosystems that occur in the otherwise nutrient poor waters of the world's tropical oceans.

However, this essential symbiosis between *Symbiodinium* and corals is sensitive to environmental stresses such as sudden changes in salinity, toxins, light levels and temperature (Hoegh-Guldberg 1999). These conditions cause the symbiosis to break down, which leads to the rapid removal of *Symbiodinium* from the tissues of the host coral. The rapid exodus of the brown *Symbiodinium* cells often results in corals turning a brilliant white colour – hence the term coral 'bleaching'. The rapid reduction in available energy resources increases the susceptibility of corals to starvation, disease and eventually death, and makes them more susceptible to competition from seaweeds and other benthic organisms (Hughes 1994; Hughes et al. 2007).

Coral bleaching was first documented almost 100 years ago, when it occurred on a small scale, particularly when intertidal corals experienced unusual tidal conditions or freshwater inundation events during rainstorms (Yonge and Nichols 1931). However, in the late 1970s, there were increased reports of entire reefs and regions experiencing simultaneous mass coral bleaching (Hoegh-Guldberg 1999). These events were first reported in the scientific literature in the early 1980s, and do not appear to have been a regular or natural response by coral reef ecosystems prior to that date. Throughout the 1980s, it also became clear that mass coral bleaching was linked to unusual 'doldrum' conditions, in which ocean temperatures exceeded long-term maxima for short periods (Glynn 1988). The association of elevated sea temperature with mass coral bleaching was confirmed in experimental studies, which show that elevating temperatures in a similar fashion would cause corals to lose their *Symbiodinium* symbionts (Hoegh-Guldberg and Smith 1989).

Satellite measurements of sea-surface temperature made regularly since the early 1980s have verified the sensitivity of reef-building corals to small elevations in sea-surface temperature (Strong et al. 1996, 2006). Subsequent studies have revealed that a 1°C increase over long-term summer maximal temperatures for a coral reef region will result in mass coral bleaching, with longer and higher sea-surface temperature anomalies leading to more widespread and intense mass coral bleaching and mortality. The success of the Hotspot programme, run by the National Oceanic and Atmospheric Administration (NOAA), has provided important verification that the major variable driving mass coral bleaching over the past three decades has been sea-surface temperature (NOAA 2012; Strong et al. 2006). However, the intensity of light during thermal stress also plays an important role in determining the extent of mass coral bleaching and mortality (Jones et al. 1998). This is primarily a function of the fact that mass coral bleaching involves a dysfunction in the ability of *Symbiodinium* to process light energy, which contributes to the production of destructive oxygen radicals such as superoxide (Hoegh-Guldberg and Jones 1999). The photoinhibition model of mass coral bleaching provides the basis for the earliest steps in the breakdown of the symbiosis between corals and *Symbiodinium* (Hoegh-Guldberg 1999).

Mass coral bleaching has affected Indo-Pacific coral reefs on a number of occasions since the early 1980s. In each of these regional and worldwide events, corals across vast areas have experienced coral bleaching, and in some cases have died in great numbers. In 1998, for example, coral reefs around Palau experienced up to 90 per cent mass bleaching, with mortality resulting in 50 per cent of heavily bleached colonies (Bruno et al. 2001). Other coral populations and reefs experienced similar conditions throughout 1998, with one estimate suggesting that the world lost 16 per cent of its corals in this single extremely warm year. Since 1998, there have been several smaller episodes of mass coral bleaching, which have had significant impacts on particular regions. Triggered by the extremely warm conditions of 2005, coral reefs in the Caribbean experienced their worst mass bleaching event on record (Eakin et al. 2009, 2010). Although scientific surveys and assessments are on-going, it appears that 2010 was a particularly difficult year for coral reefs, with mass coral bleaching occurring across the world's coral reefs. Reports from Southeast Asia indicate that large areas of reefs from Indonesia and Philippines have been impacted by mass coral bleaching in the northern summer of 2010 (Mark Eakin, coordinator of NOAA's Coral Reef Watch).

6.1.2 Impacts of ocean acidification

In addition to the impact that carbon dioxide has on ocean temperature (via the enhanced greenhouse effect), it has a profound influence on the chemistry of the ocean (Doney et al. 2009). While the consequences of a less alkaline ocean are less dramatic than those seen during a mass bleaching event, the impact of changes to ocean chemistry and pH are likely to have a lasting impact on a series of important physiological and ecological processes. Several paleo-chemical studies have revealed that oceans may not have seen pH levels as low as those occurring in today's ocean for at least a million years and possibly as long as 20 million years (Pelejero et al. 2010). Not surprisingly, there is a growing body of evidence that these changes are already having a major effect on marine species and ecosystems.

Many important coral species calcify much more slowly as the carbonate ion concentration decreases along with ocean pH (Kleypas and Langdon 2006). In this regard, experimental studies consistently show a reduction in the ability of many key coral species and other important calcifiers to precipitate their calcium carbonate skeletons. A long-term study of coral growth over the past four centuries on the Great Barrier Reef (De'ath et al. 2009) and in Thailand (Tanzil et al. 2009) has revealed that corals have been calcifying 15 per cent lower since 1990 and that previous decreases of this magnitude are not seen anywhere else in the coral records studied.

While the latter study is not definitive proof that ocean acidification alone causes a reduction in calcification and growth of corals, the broad scale at which this phenomenon is being reported suggests that the combination of elevated sea temperatures and changing carbonate chemistry has begun to affect the reef-building ability of some key coral species.

Ocean acidification also appears to affect processes other than calcification (Raven et al. 2005). A growing number of studies are revealing subtle shifts in the physiological behaviour of organisms as ocean pH decreases. Many physiological processes, such as gas exchange in fish and other organisms such as cephalopods, are extremely sensitive to shifts in ocean pH (Munday et al. 2009; Pörtner and Knust 2007; Pörtner et al. 2004). Munday and co-workers have reported that the sensory behaviour of fish is affected by small changes in ocean pH (Munday et al. 2009), with the consequence that the ability of larval fish to identify appropriate habitat and the ability of both juveniles and adults to avoid predators is severely compromised. The implications of these changes across ecosystems are complex, as a coral reef is an extremely complex environment, especially given the synergies and interactions between affected processes. Despite this complexity, it is very clear that coral reef ecosystems are responding in fundamental and complex ways as we perturb the chemical and physical characteristics of the world's oceans.

The loss of vibrant coral reef communities has extremely serious consequences for other organisms, such as fish and invertebrates, which live in and among the structure provided by reef-building corals. Given the importance of these organisms to human well-being in coastal areas, these changes, and those that appear likely to occur in the future, have serious ramifications for the future sustainability of Asia-Pacific societies.

6.2 Mangrove Ecosystems

Mangrove forests cover 137 760 km^2 of the surface of the globe, and are a prominent component of the marine ecosystems of the Asia-Pacific region, with approximately 42 per cent of the world's mangroves being found there (Spalding et al. 1997; Tabuchi 2003). The amphibious trees that make up mangrove forests are composed of a number of varieties of trees, shrubs and vines that have evolved (often independently) to live in the intertidal region between land and sea. Like coral reefs, they tend to grow along tropical coastlines, where their biodiversity and productivity are greatest. Mangroves have developed a set of complex adaptations to survive fluctuating salinities, soil anoxia, desiccation and tidal inundation (Alongi 2008; Duke et al. 1998; Hogarth 1999). Despite the physiological challenges, the high nutrient levels and solar irradiation of mangrove habitats means that primary productivity in and around mangrove forests is considerable (Alongi 2002). As a result, mangrove forests are associated with a number of highly productive fisheries based on estuarine fish and invertebrates, such as crabs and prawns. Together with fisheries, mangroves serve as a source of fuel for cooking and heating, building materials for construction, medicines, and as sites for aquaculture (Bandaranayake 1998; Baran and Hambrey 1998; Hogarth 1999; Spalding et al. 1997). In addition to these direct benefits, mangroves play incredibly important roles in protecting shorelines from direct wave action, and by trapping sediments that would otherwise be washed out to sea.

Mangroves have been heavily impacted by human activities (Waycott et al. 2009). Recent satellite assessments of mangrove forests suggest that 35 per cent of mangrove forests were lost from 1980 to 2000 (Giri et al. 2010; MA 2005). In some countries, the rate of loss was even higher. Indonesia lost half of its mangrove forests in the same period, with the inventories falling from 4.2 million to a little under 2 million hectares (Giri et al. 2010). The main reasons for these losses include unsustainable coastal agriculture and aquaculture (particularly shrimp ponds), over-exploitation of trees for fuel and construction, and dredging to make way for ports and harbours (Blasco et al. 1996; Duke et al. 2007; Kathiresan n.d.; Primavera, 2005). The combined impact of these activities is driving an overall loss of mangroves at the rate of 1–2 per cent per year.

While the associated deforestation of mangroves represents the greater near-term threat, risks to mangrove areas from sea-level rise are increasing, with the expected loss of around 10 to 20 per cent of mangroves by 2100. Sea-level rise is expected to influence the distribution of mangroves by altering the depth at which they are currently growing. In some areas, ample supplies of terrigenous material will allow the sediments in which mangroves are growing to accumulate at a rate that matches the rate of sea-level rise. Mangroves therefore might be expected to maintain their position over time in these situations. However, in many areas, sediments are unlikely to keep pace with rising sea levels and mangroves are expected to migrate in a landward direction. In many areas within the Asia-Pacific region, mangroves are hemmed in either by steep coastal geography or by human towns and infrastructure. In this situation, landward migration would not be possible and mangroves would be lost.

6.3 Seagrass Meadows

Seagrasses, derived from land plants that have re-evolved to live in aquatic settings, occupy around 177 000 km^2 of coastal areas worldwide (CI 2008). Like coral reefs and mangroves, seagrass meadows are critically important marine resources within the coastal areas of the Asia-Pacific, where they provide nurseries for a wide range of important species, stabilization of marine sediments in coastal areas, and construction materials. Seagrass meadows also provide a critically important habitat for a range of threatened (and charismatic) organisms such as dugongs and manatees. A recent global assessment has found that seagrass has been disappearing at a rate of 110 km^2 per year since 1980, and that 29 per cent of seagrass meadows have already been lost (Waycott et al. 2009). The reasons for this decline are similar to those discussed for coral reefs and mangroves, and include declining water quality (that is, increased sedimentation, pollutants and nutrient concentrations), physical destruction from coastal infilling and dredging, and changes to species through either over-exploitation of key species or the accidental introduction of alien species, such as fouling algae.

Seagrass meadows face a range of threats from global climate change including increasing temperature, sea level, storm intensity, and the direct effects of atmospheric carbon dioxide (Short and Neckles 1999). Increasing temperature is likely to affect the photosynthetic rates of seagrasses, and hence their primary productivity. The precise direction (increase or decrease) will depend on the species involved and their respective thermal tolerances (Koch et al. 2007; Zimmerman et al., 1989), which in turn is likely to affect the distribution and abundance of seagrass species (Spalding et al. 2003; Walker 1991). The associated increase in carbon dioxide will lead to increased amounts of dissolved inorganic carbon (DIC), which in turn is likely to increase the growth in biomass of some species of seagrasses (Thom 1996; Zimmerman et al. 1997). Impacts from changing temperatures will also affect reproductive behaviour, including the timing of flowering, seed release and the survival of recruits. Temperature may also influence the competitive ability of epiphytes such as microalgae, which may experience accelerated growth relative to seagrasses at higher temperatures.

The impacts of increased temperature and DIC may interact with other changes, such as increased sea level. Increasing sea level will decrease light levels in current seagrass habitats and lead to associated losses (Bjork et al. 2008; Orth et al. 2006). At the same time, as coastal areas are inundated, opportunities are likely to arise for seagrass meadows to extend landward from their current position. As with mangroves, coastal geography and/or the existence of barriers to the shoreward growth of seagrass meadows will play important roles in determining how seagrass meadows respond to changing sea levels. Patterns of coastal erosion are also likely to change as the ocean inundates coastal areas. How these factors will interact with other changes (such as intensified storm activity) is not yet clear. However, it seems likely that seagrass meadows will shift from their current positions.

6.4 Open Ocean (Pelagic) Ecosystems

In addition to coral reefs, mangroves and seagrass ecosystems, open ocean ecosystems are important in providing food and resources to people within the Asia-Pacific region. While coastal fisheries represent important and substantial resources, oceanic fisheries have offered opportunities for economic development, livelihoods and food security for countries within the Asia-Pacific region. Revenue from industries such as tuna fishing, for example, represents billions of dollars in international earnings.

Climate change is likely to affect oceanic fisheries in two ways. The first is by driving increased sea temperatures to levels currently experienced during median level El Niño events. The second is by increasing the interannual climate variability of important conditions within the open ocean waters of the Asia-Pacific. These changes may not have an equivalent today, making predictions of how they will affect pelagic fisheries such as tuna speculative. However, based on current responses of pelagic fish stocks to year-to-year variability, these changes are likely to influence the distribution and abundance of pelagic fish, and influence how economical the industry will be (McIlgorm et al. 2010; Miller 2007). Much of our insight into how these fish stocks will vary comes from their behaviour with respect to ENSO within the Asia-Pacific region.

Primary productivity in the central and eastern Pacific is likely to decline due to increased warming of the surface layers of the ocean and subsequent thermal stratification between the warmer surface waters and deeper waters. The reduction in mixing of the upper layers of the ocean with the more nutritious waters at depth appears to be already underpinning the decline in chlorophyll, and consequently ocean primary productivity, in the Pacific although the contribution of the long-term cycles to this decline has yet to be properly defined (Polovina et al. 2008). The decrease in primary productivity of the open ocean is likely to have consequences on the food webs that underpin pelagic fish stocks such as tuna (Allison et al. 2009; Badjeck et al. 2010), and lead to a spatial redistribution of fish stocks. For example, the decline in primary productivity of the central Pacific would lead to a redistribution of tuna fish stocks to high latitudes (e.g., Japan) and towards the Western equatorial Pacific. Together with increased climate variability, new approaches may well be needed for anticipating where oceanic pelagic fish stocks are to be found and harvested (Allison et al. 2009; Badjeck et al. 2010).

Long-distance fishing fleets should be able to adapt to the anticipated changes in the distribution and abundance of pelagic fish stocks. However, domestic fleets may be vulnerable to the fluctuations of tuna fisheries within their exclusive economic zones, with some countries in the central Pacific, such as Kiribati, likely to be more adversely affected than those countries in the west Pacific. States such as Kiribati, which shows a very high dependence on tuna fisheries, will ultimately be more vulnerable to the changes than those States that have less total dependence.

7. PROJECTIONS OF FUTURE CHANGE: ROLLING THE DICE

There is abundant evidence that climate change arising from the burning of fossil fuels and land-use change has already had a significant impact on marine ecosystems within the Asia-Pacific region. These changes and challenges have arisen with a relatively small change in average global temperature of around 0.8°C since the start of the Industrial Revolution. Thus, the question is raised as to what changes will arise given the current rapid pace at which greenhouse gases are being added to the Earth's atmosphere.

Global Circulation Models (GCMs) have provided important tools for understanding how human activities are likely to influence future global temperatures and climates. The Intergovernmental Panel on Climate Change (IPCC) has developed consensus across models, providing projections of future climate change that include the full range of GCMs outcomes. The IPCC model outputs use scenarios developed within the Special Report on Emission Scenarios (SRES) during the Fourth Assessment Report of the IPCC (Nakicenovic and Swart 2000; Nakicenovic et al. 2000). These SRES scenario families focus on future worlds driven economically or environmentally within a globalized or regionalized perspective. These scenario families include assumptions about economic growth, the type of energy use, and a multitude of other important variables that will determine how the world unfolds and whether greenhouse gases continue to build up in the Earth's atmosphere (Nakicenovic et al. 2000). These scenarios are highly detailed and cannot be discussed adequately here. However, the broad features of these changes are worth discussing in the context of the likely changes to marine ecosystems within the Asia-Pacific region.

The more economically focused scenarios (SRES families A1 and A2) deliver rapid economic growth in the short term, but ultimately run the risk of pushing global temperatures as high as 6.4°C above pre-industrial values. The A1 family of scenarios assumes the following: rapid economic growth, the global population reaches 9 billion by 2050 and then gradually declines, the spread of new and efficient technologies, and a convergent world in which there are extensive social and cultural interactions worldwide. Within this family, there are scenarios in which there is an emphasis on fossil fuels (A1FI: fossil fuel intensive), and others in which there is an emphasis on non-fossil fuel energy sources (A1T). The A2 scenarios assume a more divided world in which nations tend to operate independently and in a self-reliant fashion within a world that has a continuously increasing population and is divided into regionally orientated economic zones.

The more environmentally focused scenarios show a reduced range of risks. The B1 scenario family involves a more integrated, yet ecologically aware world. Rapid economic growth occurs as in A1, but rapidly transitions towards a service and information economy, with global population increasing to 9 billion and then beginning to decline as in A1. The emphasis within the B1 world is on global strategies toward achieving economic, social and environment stability. The B2 scenario family involves a more divided, yet ecologically friendly world. Population continues to increase but at a slower rate than in A2, and the emphasis is on local, rather than global, solutions to economic, social and environmental stability, with intermediate levels of economic development and less rapid and more fragmented technological change than that seen in A1 and B1.

Given the challenges that are arising with a less than 1°C change in average global temperature since the beginning of the Industrial Revolution, pursuing scenarios that minimize further changes in global temperature are desirable. In this respect, SRES scenarios belonging to the more environmentally friendly B family would be preferable. Given the strong evidence that many ecosystems, such as coral reefs, will disappear if average global temperatures exceed 2°C above the pre-industrial level, several modelling groups have recently explored what rate of emission cuts would be necessary to stabilize global temperature at or below these levels. The results of this modelling effort have produced a number of stark messages.

The first of these is that further anthropogenic emissions of carbon dioxide need to stay below a total amount of 1000 Gt CO_2 from 2000 onwards if average global temperature is to have a 75 per cent probability of remaining below 2°C. Given that global human activities release around 35 Gt CO_2 each year and 234 Gt of CO2 was released from 2000–2006, achieving this global scenario will require massive cuts in emissions over the next few decades. The widely accepted conclusion from the expert community is that emissions will have to be reduced to 80–90 per cent of today's values if we are to avoid exceeding the 2°C 'guardrail' (Richardson et al. 2009; Rockström et al. 2009).

8. CONCLUSIONS: OPPORTUNITIES FOR ADAPTATION AND COOPERATION WITHIN THE ASIA-PACIFIC REGION

Discussion of international policy imperatives and strategies to reduce emissions of greenhouse gases is beyond the scope of this chapter. There is abundant evidence that climate change arising from the burning of fossil fuels and land-use change has already had a significant impact on marine ecosystems within the Asia-Pacific region. However, one clear conclusion is drawn here for Asia-Pacific marine ecosystems based on this discussion of future scenarios; marine ecosystems will largely disappear if the trend towards atmospheric carbon dioxide concentrations of 450 ppm and global temperatures of more than 2°C above pre-industrial level is not halted. While it is hoped that the seriousness of the situation will drive nations to collectively reduce emissions to ensure the targeted levels and thus the 2°C threshold is not exceeded, it is useful to reinforce the repercussions of not cutting emissions and continuing towards rising global temperatures.

If temperatures were to rise by 2°C or more, the temperature of tropical oceans and atmospheric concentrations of carbon dioxide would increase by 1°C and another 60 ppm, respectively. Given that the concentration of carbon dioxide is growing at the rate of over 2 ppm per year, these conditions will be reached by 2040. Coral dominated reef systems will contract by as much as 50 per cent, which is in keeping with the current 1 to 2 per cent decrease in coral cover per year that has been experienced throughout the Asia-Pacific over the past 40 years (Bruno and Selig 2007). Coral reef frameworks will be more fragile and, in many cases, will not be accreting significantly. Mangroves and seagrass beds will also contract if current trends in deforestation and unsustainable coastal land-use practices continue. Ocean temperatures will increase by a further 1°C, changing the distribution of ocean ecosystems through the expansion of oligotrophic areas (Polovina et al. 2008), and through direct effects on the temperature dependent distribution of pelagic fish such as tuna (Bell et al. 2009; Miller 2007). Coastal regions within many parts of the Asia-Pacific region will experience a greater frequency of damaging storms systems, and sea levels will have risen by around 10–15 cm. The combined effect of more intense storms plus higher sea levels will mean that human communities and infrastructure in the coastal areas of the Asia-Pacific will be more vulnerable to damage. The loss of coral reef ecosystems and other coastal ecosystems such as mangroves will exacerbate this vulnerability.

As we approach these conditions within the Asia-Pacific region, several strategies become evident. First, reducing the impact of local stresses such as overfishing, coastal land degradation and marine pollution will increase the resilience and robustness of coastal ecosystems to climate impacts such as storm damage, mass coral bleaching and heat stress. Numerous studies have demonstrated that reducing other impacts on corals, mangroves and seagrass beds will increase their resilience to climate-related stresses. Secondly, in the management of fisheries, evolving, dynamic and 'adaptive' frameworks, capable of responding to changing conditions, challenges and opportunities arising from the shifting physical and chemical conditions throughout the Asia-Pacific region, will be required. Increasing the ability to monitor fisheries resources is an important part of this strategy, especially given the complex inter-relationships between climate change, local conditions and biological responses. Thirdly, greater cooperation across nations with respect to the management of pelagic fish stocks will be required, in response to changing conditions resulting in the movement of fish stocks. Establishing agreements and joint management arrangements across the Asia-Pacific will be essential to the effective management and distribution of the benefits of each of these resources. Lastly, different nations within the Asia-Pacific region are likely to be affected by climate change to different extents and will have different capabilities with respect to responding to the challenges that might arise. Therefore, cooperating across national borders and establishing regional networks of expertise and capability represents an important opportunity for responding to the enormous challenges posed by climate change. These networks will also benefit from wider international cooperation on managing the important marine resources of the Asia-Pacific, especially in regions such as the Coral Triangle, where much of the world's biodiversity resides. These wider international networks may also play important roles in directing global resources towards assisting the nations of the Asia-Pacific in finding a sustainable future for their coastal ecosystems and human communities as the current century of enormous environmental change unfolds.

REFERENCES

Allison, E., A. Perry, M. Badjeck, W. Neil Adger, K. Brown, D. Conway, A. Halls, G. Pilling, J. Reynolds and N. Andrew (2009), 'Vulnerability of national economies to the impacts of climate change on fisheries', *Fish and Fisheries*, **9999**.

Alongi, D. (2002), 'Present state and future of the world's mangrove forests', *Environmental Conservation*, **29** (3), 331–349.

Alongi, D.M. (2008), 'Mangrove forests: Resilience, protection from tsunamis, and responses to global climate change', *Estuarine Coastal and Shelf Science*, **76**, 1–13.

Andréfouët, S., C. Chauvin and C. Kranenburg (2009), *Atlas of Pacific Ocean Coral Reefs*, Centre IRD de Nouméa.

Badjeck, M.-C., E.H. Allison, A.S. Halls and N.K. Dulvy (2010), 'Impacts of climate variability and change on fishery-based livelihoods', *Marine Policy*, **34**, 375–383.

Bandaranayake, W. (1998), 'Traditional and medicinal uses of mangroves', *Mangroves and Salt Marshes*, **2**, 133–148.

Baran, E. and J. Hambrey (1998), 'Mangrove conservation and coastal management in southeast Asia: what impact on fishery resources?' *Marine Pollution Bulletin*, **37**, 431–440.

Bell, J., M. Kronen, A. Vunisea, W. Nash, G. Keeble, A. Demmke, S. Pontifex and S. Andréfouët (2009), 'Planning the use of fish for food security in the Pacific', *Marine Policy*, **33**, 64–76.

Bjork, M., E. Short, E. McLeod and S. Beer (2008), *Managing Seagrasses for Resilience to Climate Change*, IUCN Resilience Science Group Working Paper Series No. 3, Gland, Switzerland.

Blasco, F., P. Saenger and E. Janodet (1996), 'Mangroves as indicators of coastal change', *Catena*, **27**, 167–178.

Bruno, J., C. Siddon, J. Witman, P. Colin and M. Toscano (2001), 'El Nino related coral bleaching in Palau, western Caroline Islands', *Coral Reefs*, **20**, 127–136.

Bruno, J.F. and E.R. Selig (2007), 'Regional decline of coral cover in the Indo-Pacific: timing, extent, and subregional comparisons', *PLoS ONE*, **2**, e711.

Burke, L., L. Selig and M. Spalding (2002), *Reefs at Risk in Southeast Asia*, Washington DC: World Resources Institute.

Caldwell, M., T.C. Hoffmann, S. Palumbi, J. Teisch and C. Tu (2009), *Pacific Ocean: Synthesis: Scientific Literature Review of Coastal and Ocean Threats, Impacts, and Solutions*, Center for Ocean Solutions, Stanford University.

Chand, S., R.Q. Grafton and E. Petersen (2004), *Multilateral Governance of Fisheries: Management and Cooperation in the Western and Central Pacific Tuna Fisheries*, Resource Management in Asia-Pacific Region Working Technical Paper, Australian National University.

Chou, C., J.Y. Tu and J.Y. Yu (2010), 'Interannual variability of the western North Pacific summer monsoon: differences between ENSO and non-ENSO years', *Journal of Climate*, **16** (13), 2275–2287.

Church, J. and N. White (2006), 'A 20th century acceleration in global sea-level rise', *Geophysical Research Letters*, **33**.

CI (2008), *Economic Values of Coral Reefs, Mangroves, and Seagrasses: A Global Compilation*, Arlington, VA: Center for Applied Biodiversity Science, Conservation International.

Costanza, R., R. d'Arge, R. De Groot, S. Farber, M. Grasso, B. Hannon, K. Limburg, S. Naeem, R.V. O'Neill and J. Paruelo (1997), 'The value of the world's ecosystem services and natural capital', *Nature*, **387**, 253–260.

De'ath, G., J.M. Lough and K.E. Fabricius (2009), 'Declining coral calcification on the Great Barrier Reef', *Science*, **323**, 116–119.

Doney, S., V. Fabry, R. Feely and J. Kleypas (2009), Ocean acidification: the other CO2 problem', *Annual Review of Marine Science*, **1**, 169–192.

Duke, N., M. Ball and J. Ellison (1998), 'Factors influencing biodiversity and distributional gradients in mangroves', *Global Ecology and Biogeography Letters*, **7**, 27–47.

Duke, N., J. Meynecke, S. Dittmann, A. Ellison, K. Anger, U. Berger, S. Cannicci,

K. Diele, K. Ewel and C. Field (2007), 'A world without mangroves?' *Science*, **317**, 41–42.

Eakin, C.M., J.M. Lough and S.F. Heron (2009), 'Climate variability and change: monitoring data and evidence for increased coral bleaching stress', in M.J.H. Oppen and J.M. Lough (eds.), *Coral Bleaching*, Ecological Studies Series, **205**, 41–67.

Eakin, C.M., J.A. Morgan, S.F. Heron, T.B. Smith, G. Liu, L. Alvarez-Filip, B. Baca, E. Bartels, C. Bastidas and C. Bouchon (2010), 'Caribbean corals in crisis: record thermal stress, bleaching, and mortality in 2005', *PLoS ONE*, **5**, e13969.

Giri, C., E. Ochieng, L. Tieszen, Z. Zhu, A. Singh, T. Loveland, J. Masek and N. Duke (2010), 'Status and distribution of mangrove forests of the world using earth observation satellite data', *Global Ecology and Biogeography*, **20**, 154–159.

Glynn, P.W. (1988), 'El Niño warming, coral mortality and reef framework destruction by echinoid bioerosion in the Eastern Pacific', *Galaxea*, **7**, 129–160.

Griffiths, G., L. Chambers, M. Haylock, M. Manton, N. Nicholls, H.J. Baek, Y. Choi, P. Della Marta, A. Gosai and N. Iga (2005), 'Change in mean temperature as a predictor of extreme temperature change in the Asia–Pacific region', *International Journal of Climatology*, **25**, 1301–1330.

Halpern, B.S., S. Walbridge, K.A. Selkoe, C.V. Kappel, F. Micheli, C. D'Agrosa, J.F. Bruno, K.S. Casey, C. Ebert, H.E. Fox, R. Fujita, D. Heinemann, H.S. Lenihan, E.M.P. Madin, M.T. Perry, E.R. Selig, M. Spalding, R. Steneck and R. Watson (2008), 'A global map of human impact on marine ecosystems', *Science*, **319**, 948–952.

Hansen, J., R. Ruedy, M. Sato and K. Lo (2010), *Global Surface Temperature Change*, New York: NASA Goddard Institute for Space Studies, accessed 26th July 2012 at: http://pubs.giss.nasa.gov/docs/2010/2010_Hansen_etal.pdf.

Hoegh-Guldberg, O. (1999), 'Climate change, coral bleaching and the future of the world's coral reefs', *Marine and Freshwater Research*, **50**, 839–866.

Hoegh-Guldberg, O., H. Hoegh-Guldberg, J. Veron, A. Green, E.D. Gomez, A. Ambariyanto and L. Hansen (2009a), *The Coral Triangle and Climate Change: Ecosystems, People and Societies at Risk*. Sydney, NSW: WWF Australia.

Hoegh-Guldberg, O., H. Hoegh-Guldberg, J.E.N. Veron, A. Green, E.D. Gomez, J. Lough, M. King, Ambariyanto, L. Hansen, J. Cinner, G. Dews, G. Russ, H.Z. Schuttenberg, E.L. Peñafl or, C.M. Eakin, T.R.L. Christensen, M. Abbey, F. Areki, R.A., Kosaka, A., Tewfi k and J. Oliver (2009b), *The Coral Triangle and Climate Change: Ecosystems, People and Societies at Risk*, Brisbane: WWF Australia.

Hoegh-Guldberg, O. and R.J. Jones (1999), 'Photoinhibition and photoprotection in symbiotic dinoflagellates from reef-building corals', *Marine Ecology Progress Series*, **183**, 73–86.

Hoegh-Guldberg, O., P.J. Mumby, A.J. Hooten, R.S. Steneck, P. Greenfield, E. Gomez, C.D. Harvell, P.F. Sale, A.J. Edwards, K. Caldeira, N. Knowlton, C.M. Eakin, R. Iglesias-Prieto, N. Muthiga, R.H. Bradbury, A. Dubi and M.E. Hatziolos (2007), 'Coral reefs under rapid climate change and ocean acidification', *Science*, **318**, 1737–1742.

Hoegh-Guldberg, O. and G.J. Smith (1989), 'The effect of sudden changes in temperature, light and salinity on the population density and export of zooxanthellae from the reef corals *Stylophora pistillata* and *Seriatopra hystrix*', *Journal of Experimental Marine Biology and Ecology*, **129**, 279–303.

Hogarth, P.J. (1999), *The Biology of Mangroves*, Oxford, UK: Oxford University Press.
Hughes, T.P. (1994), 'Catastrophes, phase shifts, and a large-scale degradation of Caribbean coral reef', *Science*, **265**, 1547–1551.
Hughes, T.P., A.H. Baird, D.R. Bellwood, M. Card, S.R. Connolly, C. Folke, R. Grosberg, O. Hoegh-Guldberg, J.B. Jackson, J. Kleypas, J.M. Lough, P. Marshall, M. Nystrom, S.R. Palumbi, J.M. Pandolfi, B. Rosen and J. Roughgarden (2003a), 'Climate change, human impacts, and the resilience of coral reefs', *Science,* **301** (5635), 929–933.
Hughes, T.P., A.H. Baird, D.R. Bellwood, S.R. Connolly, C. Folke, R. Grosberg, O. Hoegh-Guldberg, J.B.C. Jackson, J. Kleypas, J.M. Lough, P. Marshall, M. Nystrom, S.R. Palumbi, J.M. Pandolfi, B. Rosen and J. Roughgarden (2003b), 'Causes of coral reef degradation – Response', *Science*, **302**, 1503–1504.
Hughes, T.P., M.J. Rodrigues, D.R. Bellwood, D. Ceccarelli, O. Hoegh-Guldberg, L. McCook, N. Moltschaniwskyj, M.S. Pratchett, R.S. Steneck and B. Willis (2007), 'Phase shifts, herbivory, and the resilience of coral reefs to climate change', *Current Biology*, **17**, 360–365.
IPCC (2007a), *Climate Change 2007: Impacts, Adaptation and Vulnerability. Contribution of Working Group II to the Fourth Assessment Report of the Intergovernmental Panel on Climate Change*, M.L. Parry, O.F. Canziani, J.P. Palutikof, P.J. van der Linden and C.E. Hanson (eds), Cambridge UK: Cambridge University Press.
IPCC (2007b), *Synthesis Report. Contribution of Working Groups I, II and III to the Fourth Assessment Report of the Intergovernmental Panel on Climate Change*, C.W. Team, R.K. Pachauri and A. Reisinger (eds), Geneva, Switzerland: Intergovernmental Panel on Climate Change.
Jompa, J. (2008), *Indonesian Plan of Actions: Coral Triangle Initiative on Coral Reef, Fisheries and Food Security*, Australian Workshop to Support CTI, Townsville, November.
Jones, R., O. Hoegh-Guldberg, A. Larkum and U. Schreiber (1998), 'Temperature induced bleaching of corals begins with impairment to the carbon dioxide fixation mechanism of zooxanthellae', *Plant, Cell and Environment*, **21**, 1219–1230.
Kathiresan, K. (n.d.), 'Threats to Mangroves: Degradation and desrtriction of mangroves', accessed 26th July 2012 at: http://ocw.unu.edu/international-network-on-water-environment-and-health/unu-inweh-course-1-mangroves/Degradation-and-destruction-of-mangroves.pdf.
Kitagawa, T., H. Nakata, S. Kimura, T. Itoh, S. Tsuji and A. Nitta (2000), 'Effect of ambient temperature on the vertical distribution and movement of Pacific bluefin tuna Thunnus thynnus orientalis', *Marine Ecology Progress Series*, **206**, 251–260.
Kleypas, J.A. and C. Langdon (2006), 'Coral reefs and changing seawater chemistry', in J. Phinney, O. Hoegh-Guldberg, J. Kleypas, W. Skirving and A.E. Strong (eds), *Coral Reefs and Climate Change: Science and Management*. AGU Monograph Series, Coastal and Estuarine Studies, **61**, Washington DC: Geophysics Union, 73–110.
Knutson, T.R., J.L. McBride, J. Chan, K. Emanuel, G. Holland, C. Landsea, I. Held, J.P. Kossin, A.K. Srivastava and M. Sugi (2010), 'Tropical cyclones and climate change', *Nature Geoscience*, **3**, 157–163.
Knutson, T.R., R.E. Tuleya and Y. Kurihara (1998), 'Simulated increase of hurricane intensities in a CO2-warmed climate', *Science*, **279**, 1018–1020.

Koch, M., S. Schopmeyer, C. Kyhn-Hansen, C. Madden and J. Peters (2007), 'Tropical seagrass species tolerance to hypersalinity stress', *Aquatic Botany*, **86**, 14–24.

Kompas, T. and T.N. Che (2006), 'Economic profit and optimal effort in the Western and Central Pacific tuna fisheries,' *Pacific Economic Bulletin*, **21**, 46–62.

Lehodey, P., M. Bertignac, J. Hampton, A. Lewis and J. Picaut (1997), 'El Niño Southern Oscillation and tuna in the western Pacific', *Nature*, **389**, 715–718.

Lehodey, P., F. Chai and J. Hampton (2003), 'Modelling climate related variability of tuna populations from a coupled ocean–biogeochemical populations dynamics model', *Fisheries Oceanography*, **12**, 483–494.

Levitus, S., J.I. Antonov, T.P. Boyer, R.A. Locarnini, H.E. Garcia and A.V. Mishonov (2009), 'Global ocean heat content 1955–2008 in light of recently revealed instrumentation problems', *Geophysical Research Letters*, **36**.

MA (2005), *Millennium Ecosystem Assessment. Ecosystems and Human Well-being: Synthesis*, Washington, DC: Island Press.

McIlgorm, A., S. Hanna, G. Knapp, P. Le Floc'H, F. Millerd and M. Pan (2010), 'How will climate change alter fishery governance: insights from seven international case studies', *Marine Policy*, **34**, 170–177.

Meehl, G.A., F. Zwiers, J. Evans, T. Knutson, L. Mearns and P. Whetton (2000), 'Trends in extreme weather and climate events: issues related to modeling extremes in projections of future climate change', *Bulletin of the American Meteorological Society*, **81**, 427–436.

Miller, K.A. (2007), 'Climate variability and tropical tuna: management challenges for highly migratory fish stocks', *Marine Policy*, **31**, 56–70.

Moberg, F. and C. Folke (1999), 'Ecological goods and services of coral reef ecosystems', *Ecological Economics*, **29**, 215–233.

Moberg, F. and P. Rönnbäck (2003), 'Ecosystem services of the tropical seascape: interactions, substitutions and restoration', *Ocean and Coastal Management*, **46**, 27–46.

Munday, P., D. Dixson, J. Donelson, G. Jones, M. Pratchett, G. Devitsina and K. Døving (2009), 'Ocean acidification impairs olfactory discrimination and homing ability of a marine fish', *Proceedings of the National Academy of Sciences*, **106**, 1848.

Muscatine, L. (1990), 'The role of symbiotic algae in carbon and energy flux in reef corals', in Z. Dubinsky (ed.), *Ecosystems of the World: Coral Reefs*, Amsterdam: Elsevier, pp. 75–87.

Nakicenovic, N., J. Alcamo, G. Davis, B. de Vries, J. Fenhann, S. Gaffin, K. Gregory, A. Grubler, T. Jung and T. Kram (2000), *Special Report on Emissions Scenarios: a Special Report of Working Group III of the Intergovernmental Panel on Climate Change*, PNNL-SA-39650, New York: Cambridge University Press.

Nakicenovic, N. and R. Swart (2000), *Emissions Scenarios 2000 – Special Report of the Intergovernmental Panel on Climate Change*, Cambridge, UK: Cambridge University Press, **96**, 98.

NOAA (2012), 'Coral Reef Watch satellite monitoring', accessed 26th July 2012 at http://coralreefwatch.noaa.gov/satellite/index.html.

Orth, R.J., T. Carruthers, W.C. Dennison, C.M. Duarte, J.W. Fourqurean, K.L Heck Jr, A.R. Hughes, G.A. Kendrick, W.J. Kenworthy and S. Olyarnik (2006), 'A global crisis for seagrass ecosystems', *BioScience*, **56**, 987–996.

Pelejero, C., E. Calvo and O. Hoegh-Guldberg (2010), 'Paleo-perspectives on ocean acidification', *Trends in Ecology and Evolution*, **25** (6), 332–344.

Peñaflor, E., W. Skirving, A. Strong, S. Heron and L. David (2009), 'Sea-surface temperature and thermal stress in the Coral Triangle over the past two decades', *Coral Reefs*, **28**, 841–850.

Polovina, J., E. Howell and M. Abecassis (2008), 'Ocean's least productive waters are expanding', *Geophysical Research Letters*, **35**, L03618.

Pörtner, H. and R. Knust (2007), 'Climate change affects marine fishes through the oxygen limitation of thermal tolerance', *Science*, **315**, 95.

Pörtner, H., M. Langenbuch and A. Reipschläger (2004), 'Biological impact of elevated ocean CO2 concentrations: lessons from animal physiology and earth history', *Journal of Oceanography*, **60**, 705–718.

Pratchett, M., P.L. Munday, N.A.J. Graham, M. Kronen, S. Pinca, K. Friedman, T. Brewer, J.D. Bell, S.K. Wilson, J.E. Cinner, J.P. Kinch, R.J. Lawton, A.J. Williams, L. Chapman, F. Magron and A. Webb (2011), 'Coastal fisheries', in J. Bell (ed.), *Vulnerability of Fisheries and Aquaculture in the Pacific to Climate Change*, Suva, Fiji: South Pacific Commission.

Primavera, J.H. (2005), 'Global voices of science: mangroves, fishponds, and the quest for sustainability', *Science*, **310**, 57–59.

Raven, J., K. Caldeira, H. Elderfield, O. Hoegh-Guldberg, P. Liss, U. Riebesell, J. Shepherd, C. Turley and A. Watson (2005), *Ocean Acidification due to Increasing Atmospheric Carbon Dioxide*, Policy document 12/05.

Reaka-Kudla, M.L. (1997), 'Global biodiversity of coral reefs: a comparison with rainforests', in M.L. Reaka-Kudla and D.E. Wilson (eds), *Biodiversity II: Understanding and Protecting our Biological Resources, Volume II*, Joseph Henry Press, p. 551.

Richardson, K., W. Steffen, H.J. Schellnhuber, J. Alcamo, T. Barker, D.M. Kammen, R. Leemans, D. Liverman, M. Munasinghe, B. Osman-Elasha, N. Stern and O. Wæver (2009), *Synthesis Report, Climate Change: Global Risks, Challenges and Decisions*, Climate Congress, Copenhagen 10–12 March, accessed 26th July 2012 at: http://climatecongress.ku.dk/.

Rockström, J., W. Steffen, K. Noone, A. Persson, F. Chapin, E. Lambin, T. Lenton, M. Scheffer, C. Folke and H. Schellnhuber (2009), 'A safe operating space for humanity', *Nature*, **461**, 472–475.

Ronnback, D. (1999), 'The ecological basis for economic valuation of seafood production supported by mangrove ecosystems', *Ecological Economics*, **29**, 235–252.

Short, F.T. and H.A. Neckles (1999), 'The effects of global climate change on seagrasses', *Aquatic Botany*, **63**, 169–196.

Spalding, M., F. Blasco and C. Field (1997), *World Mangrove Atlas*, Okinawa, Japan.

Spalding, M., H. Fox, G. Allen, N. Davidson, Z. Ferdana, M. Finlayson, B. Halpern, M. Jorge, A. Lombana and S. Lourie (2007), 'Marine ecoregions of the world: a bioregionalization of coastal and shelf areas', *BioScience*, **57**, 573–583.

Spalding, M., C. Ravilious and E. Green (2001), *World Atlas of Coral Reefs*, Berkeley: University of California Press.

Spalding, M., M. Taylor, C. Ravilious, F. Short and E. Green (2003), 'The distribution and status of seagrasses', in E. Green and F. Short (eds), *World Atlas of Seagrasses*, Berkeley: University of California Press, pp. 5–26.

Stowasser, M., Y. Wang and K. Hamilton (2010), 'Tropical cyclone changes in the western North Pacific in a global warming scenario', *Journal of Climate*, **20**, 2378–2396.

Strong, A., F. Arzayus, W. Skirving and S. Heron (2006), 'Identifying coral bleaching remotely via Coral Reef Watch – improved integration and implications for changing climate', in J.T. Phinney et al. (eds), *Coral Reefs and Climate Change: Science and Management*, Coastal and Estuarine Studies Series, **61**, pp. 163–180.

Strong, A.E., C.S. Barrientos, C. Duda and J. Sapper (1996), *Improved Satellite Technique for Monitoring Coral Reef Bleaching*, Proceedings of the Eighth International Coral Reef Symposium, Panama, June, pp. 1495–1497.

Tabuchi, R. (2003), 'The rehabilitation of mangroves in Southeast Asia', accessed 26th July 2012 at: http://info.frim.gov.my/cfdocs/tsonami2/The%20rehabilitation%20of%20mangroves%20in%20Southeast%20Asia%20-%20TabuchiCH12.pdf.

Tanzil, J., B. Brown, A.Tudhope and R. Dunne (2009), 'Decline in skeletal growth of the coral Porites lutea from the Andaman Sea, South Thailand between 1984 and 2005', *Coral Reefs*, DOI 10.1007/s00338-008-0457-5.

Thom, R. (1996), 'CO2-Enrichment effects on eelgrass (Zostera marina L.) and bull kelp (Nereocystis luetkeana (mert.) P & R', *Water, Air and Soil Pollution*, **88**, 383–391.

Walker, D. (1991), 'The effect of sea temperature on seagrasses and algae on the Western Australian coastline', *Journal of the Royal Society of Western Australia*, **74**, 71–77.

Wang, M.Y. and J.E. Overland (2009), 'A sea ice free summer Arctic within 30 years?' *Geophysical Research Letters*, **36**, L07502.

Wang, M.Y., J.E. Overland, N. Bond and A.M.S. Ams (2002), *The Climate Change in the Western Arctic During the Last Two Decades*, 13th Symposium on Global Change and Climate Variations, Orlando, FL, pp. 40–41.

Waycott, M., C.M. Duarte, T.J.B. Carruthers, R.J. Orth, W.C. Dennison, S. Olyarnik, A. Calladine, J.W. Fourqurean, K.L. Heck, A.R. Hughes, G.A. Kendrick, W.J. Kenworthy, F.T. Short and S.L. Williams (2009), 'Accelerating loss of seagrasses across the globe threatens coastal ecosystems', *Proceedings of the National Academy of Sciences of the United States of America*, **106**, 12377–12381.

Woodward, A., S. Hales and P. Weinstein (1998), 'Climate change and human health in the Asia Pacific region: who will be most vulnerable?' *Climate Research*, **11**, 31–38.

Yonge, C.M. and A.G. Nichols (1931), 'Studies on the physiology of corals: V. The effect of starvation in light and in darkness on the relationship between corals and zooxanthellae', *Scientific Report of the Great Barrier Reef Expedition*, **1**, 177–211.

Zimmerman, R., D. Kohrs, D. Steller and R. Alberte (1997), 'Impacts of CO_2 enrichment on productivity and light requirements of eelgrass', *American Society of Plant Biologists*, **115**, 599–607.

Zimmerman, R., R. Smith and R. Alberte (1989), 'Thermal acclimation and whole-plant carbon balance in *Zostera marina* L. (eelgrass)', *Journal of Experimental Marine Biology and Ecology*, **130**, 93–109.

3. Uncertain seas ahead: legal and policy approaches to conserving marine biodiversity in the face of changing climate

Richard Kenchington and Robin Warner

1. INTRODUCTION

Climate is a major factor in the habitat, food chains, competition, success and survival of species. Contemporary distributions and abundance of marine species and communities reflect adaptation to geologically recent climatic conditions and the impacts of human activities. Warming of the atmosphere and seawater has occurred in association with increasing levels of atmospheric carbon dioxide since the start of the twentieth century. Despite continuing scientific research and wider discussion of the relative roles of anthropogenic greenhouse gas increases and other influences on climate, climate change is occurring. The policy and legal issues have two core components: response to the effects of climate change, and addressing the human activities for which there is reasonable evidence of causation or exacerbation of climate change. For the purpose of this chapter, the focus will be on the response to the effects of climate change, rather than on the issue of anthropogenic causation and exacerbation.

The effects of climate change on marine biodiversity flow from increasing water temperature and absorption of carbon dioxide from the atmosphere with consequential changes in the chemistry of seawater; the strength and direction of ocean currents; and the intensity, frequency and geographic range of extreme weather events. The expected consequences of recent and projected anthropogenic increases in greenhouse gases on climate change are now considered inevitable, with temperatures set to continue to increase. This is because the period over which any stabilization or return to historic levels would occur is expected to be long.

In policy and legal terms, the effects of climate change on marine biodiversity compound and are difficult to separate from the effects caused by anthropogenic impacts such as the overexploitation of fisheries and marine resources; coastal habitat destruction; and operational and catastrophic accidental pollution arising from marine industries, shipping and land and freshwater uses. The combined effects may be linked over substantial distances, within and between jurisdictions, by run-off from land, and by currents transporting larvae, nutrients and food in water columns.

This raises issues that require multisectoral integration of policy and management within jurisdictions, coordination with adjacent and linked jurisdictions and regional and international mechanisms to address areas beyond national jurisdictions. These issues have been matters of concern since the United Nations (UN) Conference on the Human Environment in Stockholm in 1972. This conference led to the establishment of the United Nations Environment Programme (UNEP) and subsequently to the World Conservation Strategy (1987) and the World Environment Summit in Rio de Janeiro in 1992, which in turn produced in the UN Convention on Biological Diversity and Agenda 21. Chapter 17 of Agenda 21 specifically addresses the protection of oceans, seas (including enclosed and semi-enclosed seas) and coastal areas, as well as the protection, rational use and development of their living resources.

The complexities of multisectoral and trans-jurisdictional policy and management have led to many systems and approaches designed to address specific sets of circumstances; for example, Integrated Coastal Zone Management, Integrated Coast and Ocean Management, Marine Spatial Planning and Ecosystem-based Management.

A World Bank (2006, pp. 9–12) report listed 32 marine management tools and developed a typology based on the objectives and extent of the environmental protection offered. It identified four groups:

- marine protected area tools, primarily for biodiversity conservation and habitat protection
- multiuse management tools, primarily for balanced conservation and socioeconomic uses
- sustainable use marine-resource management tools, primarily for extractive use
- culture/ecological/social protection reserves, primarily for indigenous and traditional non-indigenous communities.

This list is not exhaustive, but it reflects the social and political challenge of integrating the objectives of competing sectoral approaches and addressing overarching issues such as the predicted effects of climate-related changes and the increasing range of human uses and impacts affecting marine space.

2. CLIMATE CHANGE IMPACTS ON MARINE BIODIVERSITY

Changes in which climate is a major or significant driver include: increasing water temperature; changes in the chemical properties of seawater; sea level rise; increased frequency, severity and range of severe weather events; and increasing thermal and other stresses on species due to all these factors. These changes are discussed in more detail below.

2.1 Increasing Water Temperature

A core consideration in the adaptation of a species or individual to changing climate is the thermal tolerance range of that species, defined by its upper and lower lethal temperature limit, and, within that, the optimal thermal range. Deser et al. (2010) reviewed tropical sea-surface temperature trends for the twentieth century. Their analysis was limited by poor and inconsistent historic sampling and measurement practices. However, they found reasonable concurrence in estimates of 0.35°C for the century. For Australian regional seawater surface temperature rises to 2070, the predicted range is from 0.6°C to 2.5°C (Commonwealth of Australia 2007). The expected consequences of ocean warming include increased thermal stress on tropical species and communities, with the probability of substantial change and significant species loss (Cheung et al. 2009).

Typically, the distribution of tropical species reflects a range close to absolute upper level of thermal tolerance. Under increasing temperatures, they are likely to show local extinction in their original habitats and, where possible, movement or invasion through larval transport and settlement to follow the movement of their preferred thermal range to higher latitudes. Some or many species may survive by gradual relocation through range extension or larval settlement in suitable habitats in higher latitude areas that have warmed. This is likely to cause a cascade effect, with such relocation causing competition with and displacement of species for which the temperature range of the invaded location has become too high. At polar latitudes, the pressures on species adapted to extremes of cold environmental conditions are expected to lead to species loss.

Cheung et al. (2009) have modelled likely patterns of such changes and predicted significant species extirpation in the equatorial South Asian/Indian Ocean and west Pacific Ocean, with the highest levels in the enclosed Java Sea, but low levels of species invasion in the Asia-Pacific region generally. They did not address the potential implications of climate change on habitat-building species such as corals, so the predicted levels of extirpation may be regarded as conservative.

The immediate policy and legal issues arising from species distribution changes are likely to relate to species of fisheries importance. Cheung et al. (2009) modelled projected changes to 2055 in maximum fishery catch potential. Their projections show widespread reductions of 30 to 50 per cent or more in most of the equatorial Asia-Pacific region through reductions in current tropical fisheries. Conversely, they predict increases in excess of 100 per cent in the south-eastern tropical Indian Ocean and sub-tropical and temperate south-west Pacific.

In addition to the changes expected to flow from gradually rising mean sea-surface temperatures, relatively short periods of extreme temperature rise can also have major ecological consequences. In 1998/99, substantial areas of the Indo-Pacific experienced a prolonged period of severe high water temperature, which caused widespread coral bleaching. Many days of water temperatures two or more degrees above normal summer maxima stressed corals, which responded by rejecting their symbiotic algae, zooxanthellae. The stress continued for so long that there was widespread coral death, with large areas experiencing more than 95 per cent coral mortality. In many of these areas, there has been recovery of coral cover through recruitment. However, recovery of the structural complexity and requisite habitat for the broad range of species associated with healthy coral reefs is a longer-term prospect.

2.2 Changes in the Chemical Properties of Seawater

Seawater is a complex and dynamic solution of interacting salts and ions that interact directly with the atmosphere at the sea surface. Atmospheric gases dissolve in seawater to an extent determined by their partial pressure and the consequential chemical reactions they have with other solutes. The increasing levels of atmospheric carbon dioxide have translated into increasing concentrations of carbon dioxide in the oceans, causing the acidification of seawater. The consequences are difficult to predict because of the complex dynamics of seawater chemistry, but there is evidence that acidification is reducing the density of calcium carbonate in the skeletons of corals and other calcifying species, including planktonic species.

2.3 Sea-level Rise

International Panel on Climate Change (IPCC) model-based predictions of sea-level rise for the remainder of the twenty-first century are 20 to 42 mm per decade in a low scenario and 28 to 65mm per decade in a high scenario. These scenarios exclude considerations of ice flow because of the lack of published literature. Douglas (1997) used long time-series data from tide gauges to derive a global mean rise of sea level of 20 mm per decade for the twentieth century. There are regional differences and Webb (2010) discusses Pacific Regional Island Shoreline Monitoring System data from studies of 27 islands for between 20 and 60 years, showing island stability and some areas of land increase, with a mean sea level increase of 100 mm since the mid-twentieth century.

The phenomenon of sea level rise is real, but it is occurring amidst tidal cycles and the greater shorter-term variations caused by the El Niño Southern Oscillation (up to 600 mm over periods of months every 5–10 years) and atmospheric pressure (from 700 to 1300 mm over periods of hours to months).

The significance for biodiversity is that the effects of sea-level rise will include inundation of low-lying areas and changes to the availability of habitats for intertidal and sub-tidal plants and animals. In unpopulated and undeveloped areas, the habitats of sub-tidal and intertidal plant and animal communities would follow the gradual landward movement of the tidal band. Elsewhere it can be expected that actions to protect property and infrastructure will restrict this process.

2.4 Increasing Frequency and Intensity of Severe Weather Events

While warming may be a gradual process, the changes it brings are expected to be delivered through weather events including severe destructive cyclonic storms and extremes of temperature, drought and rainfall. Before mean sea level reaches levels predicted in IPCC scenarios, severe storm surge events beyond high water are likely to have more frequent major destructive impacts on beaches, islands, low lying land, urban, industrial, agricultural and communications infrastructure, and on soils that are increasingly flooded with seawater.

2.5 Issues of Scale in Space and Time

The significance of climate change for biodiversity is that animal and plant communities are shaped by severe events. Shallow marine and coastal communities, particularly those in exposed coastal areas, are directly and quite frequently impacted by events such as severe storm waves, extremes of temperature and freshwater dilution or displacement of seawater through flooded river run-off plumes or intense localized rainfall events. Such events cause substantial damage, death or removal of species in an impacted area, but they also provide subsequent opportunities for recruitment or invasion of plants and animals to colonize that affected area.

Initial recovery after such events may involve regeneration of surviving damaged benthic species such as sea grasses and corals; larval recruitment of opportunistic short-lived species different to those that were displaced; and, over time, subsequent succession may lead to different mature communities from those destroyed by the impact event. Connell, Hughes and Wallace (1997) discuss long-term variations observed in 30 years of study of corals on Heron Island Reef and highlight that the mechanisms that influence abundance operate over many scales of space and time, with the consequence that studies on small and large scales are needed to understand them.

The critical factor for future policy and management is the expectation that gradual changes in sea level, seawater temperature and chemistry will be accompanied by an increased frequency of intense events. Recovery intervals for affected biological communities are consequently expected to reduce, favouring resilient, fast growing and opportunistic species. Long-term studies, such as that reported by Connell, Hughes and Wallace (1997), are rare and this limits current capacity to predict the changes likely to occur in biological communities and their implications.

3. POLICY AND LEGAL IMPLICATIONS OF CLIMATE CHANGE FOR BIODIVERSITY

The effects of climate change will increase the stress on plant and animal communities and the ecological services they provide. These effects compound stresses from human uses such as pollution, habitat loss and damage, and changed environmental flows of freshwater, sand and nutrients.

It is clear from experience of extreme events, such as coral bleaching and severe storm impacts, that the likely effects of climate change will not be uniformly distributed. The life cycles of many of the species on the seabed or in the water column of a specified area may include breeding sites, larval growth areas and migrations outside and often far beyond the boundaries of a specified management area. The economic benefits derived in one place in a catchment may have significant environmental and economic costs at another location lower in the same catchment or in coastal waters where a river reaches the sea.

Areas remote from significant centres of human population or industrial activity, and with intact biological communities and predator prey components, are likely to be less impacted. For other locations, an important issue is resilience: the capacity to survive and return to normal functional efficiency following a departure from preferred range conditions. In part, this reflects the health of the communities and, in part, can reflect connectivity to areas that have been less affected.

Understanding and managing human uses on a basis of verifiable sustainability, and the identification and protection of suitable areas as reference sites and sanctuaries becomes particularly important in the face of climate change. Reference sites enable the monitoring and understanding of changes, while sanctuaries provide for mature breeding populations whose offspring can re-populate impacted areas. The maintenance of biodiversity in its broadest sense, and the support of fisheries and other ecosystem services are likely to become an increasingly important issue of food and resource security.

The policy options for managing marine biodiversity relate primarily to achieving a sustainable balance of measures that minimize direct anthropogenic stresses on the capacity of species and communities to survive and adapt in the face of climate change, human uses and impacts, and the on-going natural biophysical dynamics affecting marine ecosystems. This requires an approach to policy and the management of human activities and impacts within the constraints of ecosystem function. A current term for such an approach is 'ecosystem-based management', which is described by the UNEP as:

> In ecosystem-based management, the associated human population and economic/social systems are seen as integral parts of the ecosystem. Most importantly, ecosystem-based management is concerned with the processes of change within living systems and sustaining the goods and services that healthy ecosystems produce. Ecosystem-based management is therefore designed and executed as an adaptive, learning-based process that applies the principles of the scientific method to the processes of management. (UNEP 2011, p. 13)

The core of the approach is coordination of sectoral management within an institutional arrangement that establishes and addresses mutually sustainable ecological, social and economic objectives and constraints. Within a jurisdiction, this may be addressed by overarching legislation or policy that integrates and binds agencies within an operational framework of explicit objectives. Between jurisdictions, such an arrangement may be addressed through agreement to coordinate policy and operations within a framework of policy objectives (Kenchington and Crawford 1993). The task of marine ecosystem management is made particularly challenging by the combination of climate and other far-reaching changes. Many of the elements that have to be addressed in management are strongly connected across jurisdictional and sectoral boundaries.

4. LEGAL AND POLICY FRAMEWORKS FOR CONSERVATION OF MARINE BIODIVERSITY AND LINKAGES WITH CLIMATE CHANGE

Scientists acknowledge that measures to conserve marine biodiversity are important bulwarks against the adverse effects of climate change on marine species, their habitats and ecosystem structures. Even before the emergence of climate change as a driving issue, global, regional and national communities had invested significant efforts in establishing legal and policy frameworks to support the conservation of terrestrial and marine biodiversity. The World Congress on National Parks in 1962 was one of the first international conservation meetings to address marine management from an ecosystem conservation perspective. The need for a systematic approach to establishing protected areas in marine environments was first clearly articulated at an International Conference on Marine Parks and Protected Areas, convened in Tokyo in 1975 by the International Union for Conservation of Nature (IUCN 1976). This concept was revisited in 1988 at the Seventeenth General Assembly of the IUCN, which recognized the urgency of the need for a spectrum of measures addressing the roles of conservation, and adopted a resolution with the primary goal:

> To provide for the protection, restoration, wise use, understanding and enjoyment of the marine heritage of the world in perpetuity through the creation of a global representative system of marine protected areas and through the management, in accordance with the principles of the World Conservation Strategy, of human activities that use or affect the marine environment. (IUCN 1988, p. 105, para. 17.38)

In the narrower sense, this meant the strict protection of special areas, while in the broader sense, it meant sustainability and stewardship consistent with the *World Conservation Strategy* (IUCN/UNEP/WWF 1980).

Key elements and mechanisms within these frameworks provide the legal authority for recognizing the adverse effects of climate change on marine biodiversity and taking remedial measures to adapt and mitigate its worst effects. At the global level, the 1982 United Nations Convention on the Law of the Sea (LOSC) signalled the advent of a more holistic approach to the protection of the marine environment. Article 194(5) recognized that measures taken to protect and preserve the marine environment should include those necessary to protect and preserve rare or fragile ecosystems, as well as the habitat of depleted, threatened or endangered species and other forms of marine life. A decade later, the 1992 Convention on Biological Diversity (CBD) was negotiated as a conventional international law framework to assist States in arresting the alarming rate of extinction of species and destruction of their habitats (Birnie et al. 2009, pp. 612–613; Grubb et al. 1993, p. 75; Joyner 1995, p. 644). The provisions of the CBD share similarities with the vision enunciated in the Rio Declaration and Agenda 21, as agreed upon at the United Nations Convention on Environment and Development (UNCED), of integrated and ecosystem-based management of the environment, including of marine areas beyond national jurisdictions (Grubb et al. 1993, pp. 75–76). The three broad objectives of the CBD, set out in Article 1, are the conservation of biodiversity, the sustainable use of its components, and the fair and equitable sharing of the benefits arising out of the utilization of genetic resources. Chapter 17 of Agenda 21 concerns the protection of the oceans, seas (including enclosed and semi-enclosed seas), and coastal areas, and the protection, rational use and development of their living resources. It identifies specific needs for marine conservation; calling for new approaches to marine and coastal area management and development at the national, subregional, regional and global levels. It recommends that these approaches be integrated in content and precautionary and anticipatory in ambit, as reflected in the following programme areas:

a. integrated management and sustainable development of coastal areas, including exclusive economic zones
b. marine environmental protection
c. sustainable use and conservation of marine living resources of the high seas
d. sustainable use and conservation of marine living resources under national jurisdiction
e. addressing critical uncertainties for the management of the marine environment and climate change
f. strengthening international, including regional, cooperation and coordination
g. sustainable development of small islands.

In support of these objectives, the Contracting Parties have developed a variety of supplementary guidelines that elaborate on key tools for mitigating the adverse effects of human activities on biodiversity. These include environmental impact assessments (EIA), strategic environmental assessments and marine spatial planning. At the Asia-Pacific regional level, specific legal and policy frameworks to conserve marine biodiversity and reduce the negative effects of human activity on regional marine ecosystems have emerged through organizations and arrangements such as Partnerships in Environmental Management for the Seas of East Asia

(PEMSEA), the Coral Triangle Initiative (CTI) and the South Pacific Regional Marine Environment Programme (SPREP). These include some specific initiatives to monitor the impacts of climate change on regional marine ecosystems and address efforts towards alleviating adverse effects.

At the national level, enactments such as Australia's 1975 Great Barrier Reef Marine Park Act and 1999 Environment Protection and Biodiversity Conservation Act (EPBC) incorporate the processes necessary to identify the adverse effects of human activity on marine biodiversity, including the effects linked to anthropogenically induced climate change. Further, these Acts provide the authority to impose relevant mitigation measures on the perpetrators of such activities. Australia's marine bioregional planning process is also taking into account the projected impacts of climate change on coastal and offshore areas under national jurisdiction.

The following section will further examine some global, regional and national legal frameworks and initiatives for the conservation of marine biodiversity to determine how capable they are of recognizing climate change impacts and limiting their adverse effects.

4.1 Global Frameworks

The LOSC established a spatially based framework of jurisdictional rights and responsibilities for the management of living resources and the protection and preservation of the marine environment. Table 3.1 sets out key provisions of the LOSC relevant to conservation and management of marine living resources and protection and preservation of the marine environment.

4.1.1 1992 Convention on Biological Diversity

The CBD provides a set of guiding principles and recommended processes for Contracting Parties establishing national programmes for biodiversity conservation. Moreover, it highlights the need for *in situ* conservation of biodiversity, which is defined in Article 2 as 'the conservation of ecosystems and natural habitats and the maintenance and recovery of viable populations of species in their natural surroundings' (Grubb et al. 1993, pp. 82–83; Kimball 1995, p. 765). These elements can also be applied in any programme implemented collaboratively by States to conserve marine biodiversity across marine boundaries and in marine areas beyond national jurisdiction. Under Article 7, Contracting Parties are directed to identify components of biodiversity important for conservation and sustainable use, with an indicative list of categories set out in Annex I. Following identification, Contracting Parties are advised to monitor, through sampling and other techniques, these components of biodiversity,

Table 3.1 Provisions of the United Nations Convention on the Law of the Sea particularly relevant to the management of living resources and the protection and preservation of the marine environment

Part V Exclusive Economic	Article 61:	Conservation of living resources
	Article 62:	Utilization of living resources, including provisions for access to resources not utilized by the coastal state
Part VII High Seas	Article 118:	Cooperation of States in the conservation and management of living resources
	Article 119:	Conservation of the living resources of the high seas
	Article 120:	Marine mammals
Part XII Protection and Preservation of the Marine Environment	Article 192:	General obligation to protect and preserve the marine environment
	Article 194:	Measures to prevent, reduce and control pollution of the marine environment
	Article 197:	Cooperation on a global or regional basis
	Article 206:	Assessment of potential effects of activities
	Article 237:	Obligations under other conventions on the protection and preservation of the marine environment

paying particular attention to the need for urgent conservation measures and to those components that offer the greatest potential for sustainable use.

As part of these initial steps towards biodiversity conservation, Contracting Parties are advised to identify processes and categories of activities that have or are likely to have significant adverse impacts on the conservation and sustainable use of biodiversity, and to monitor their effects. Data obtained from these identification and monitoring processes are to be maintained and organized by Contracting Parties. This process of information gathering specified in the CBD provisions is relevant to climate change, as it will capture information related to climate change impacts on marine ecosystems and data on human activities causally linked to climate change, such as the emission of greenhouse gases by industry.

Two key biodiversity conservation measures outlined in Articles 8 and 14 of the CBD are closely linked to identifying and mitigating climate change impacts on marine biodiversity. Article 8 provides a comprehensive description of the principles and measures associated with *in situ* conservation of biodiversity, advising Contracting Parties to promote the protection of ecosystems, natural habitats and the maintenance of viable populations of species in natural surroundings, to rehabilitate and restore degraded ecosystems and to promote the recovery of threatened species. One of the principal means of achieving *in situ* conservation emphasized in the CBD is the establishment of protected areas or areas in which special measures need to be taken to conserve biodiversity. Under Article 8(b), Contracting Parties are directed to develop guidelines for the selection, establishment and management of such areas. The processes already implemented by States at the national level to identify and manage marine protected areas or areas in which special measures are taken to conserve marine biodiversity can also be utilized to capture and monitor information on climate change impacts and to introduce mitigation measures.

Under Article 14 of the CBD, Contracting Parties are advised to introduce EIA procedures for proposed projects that are likely to have significant adverse effects on biodiversity, to avoid or minimize such effects. They are also urged to promote notification, exchange of information and consultation on activities under their jurisdiction or control that are likely to have significant adverse impacts on the biodiversity of areas beyond national jurisdiction by encouraging the conclusion of regional and multilateral arrangements. The introduction of EIA processes at national and regional level provides a further means of capturing information on climate change impacts on marine biodiversity and developing mitigation measures to address their adverse effects.

The Conference of the Parties to the CBD (COP CBD) has established an Ad Hoc Technical Expert Group on Biodiversity and Climate Change. At its tenth meeting in October 2010, the COP CBD considered the findings of the Second Ad Hoc Technical Expert Group on Biodiversity and Climate Change and recommended that Contracting Parties and other Governments consider the guidance of the Committee on a range of matters relating to biodiversity and climate change (CBD 2012, X/33 para. 1). In particular, they recommended that States identify, monitor and address the impacts of climate change and ocean acidification on biodiversity and ecosystem services and assess the future risks for biodiversity and the provision of ecosystem services using the latest available vulnerability and impact assessment frameworks (CBD 2012, X/33 para. 8(a)). They also recommended a number of strategies, some of which are particularly relevant to marine biodiversity, to reduce the impacts of climate change on biodiversity and increase the adaptive capacity of species and the resilience of ecosystems in the face of climate change. These strategies include:

- reducing non-climatic stresses such as pollution, over-exploitation, habitat loss and fragmentation, and invasive alien species
- reducing climate-related stresses where possible, through enhanced adaptive and integrated marine and coastal management
- strengthening protected area networks
- integrating biodiversity into wider seascape and landscape management
- restoring degraded ecosystems and ecosystem functions
- facilitating adaptive management by strengthening monitoring and evaluation systems (CBD 2012, X/33 para. 8(d)).

COP 10 recommended that States develop a strategy for biodiversity conservation and sustainable use that includes seascape management in those areas that are becoming accessible to new uses as a consequence of climate change, and that specific measures be taken for species that are vulnerable to climate change, including migratory species (CBD 2012, X/33 para. 8(f), (g), (i)). Ecosystem-based approaches for climate change mitigation were suggested, including enhancing the conservation, sustainable use and restoration of marine and coastal habitats that are vulnerable to the effects of climate change or which contribute to climate change mitigation, such as mangroves, peatlands, tidal salt marshes, kelp forests and seagrass beds (CBD 2012, X/33 para. 8(j), (m)).

In relation to climate change mitigation and adaptation measures, COP 10 emphasized the need to take into account the effects of such activities on marine biodiversity and the provision of ecosystem services through building on a scientifically credible knowledge base and developing ecosystem and species vulnerability assessments (CBD 2012, X/33 para. 8(v)). In particular, States were urged to ensure that no climate-related geo-engineering activities, such as ocean fertilization, that may affect biodiversity take place until there is an adequate scientific basis to justify such activities. Further, appropriate consideration of the associated risks to the environment and biodiversity was encouraged (CBD 2012, X/33 para. 8(w)).

In its decisions on marine and coastal biodiversity, COP 10 highlighted the adverse impact of climate change on marine and coastal biodiversity and recognized that the ocean is one of the largest natural reservoirs of carbon, which can significantly affect the rate and scale of global climate change (CBD 2012, X/29 para. 7). COP 10 expressed serious concern that increasing ocean acidification, as a direct consequence of increased carbon dioxide concentration in the atmosphere, reduces the availability of carbonate minerals in seawater, which are important building blocks for marine plants and animals. Therefore, it was recommended that the ecological effects of ocean acidification be considered in conjunction with the impacts of global climate change (CBD 2012, X/29 para. 64). To this end, COP 10 proposed that the CBD develop a series of joint expert review processes to monitor and assess the impacts of ocean acidification on marine and coastal biodiversity in collaboration with other international organizations including the Intergovernmental Oceanographic Commission of the United Nations Educational, Scientific and Cultural Organization (IOC/UNESCO), the Food and Agriculture Organization (FAO), the Secretariat of the United Nations Framework Convention on Climate Change (UNFCCC), the World Conservation Monitoring Centre of the United Nations Environment Programme (UNEP-WCMC), the International Coral Reef Initiative (ICRI), the Ramsar Convention, the Antarctic Treaty, and the Arctic Council, and that the results of these assessments be transmitted to the UNFCCC Secretariat (CBD 2012, X/29 para. 66).

The COP 10 decision on marine and coastal biodiversity placed particular emphasis on the application of the scientific criteria developed by the CBD COP 9 for the identification of ecologically and biologically significant areas (EBSAs). These provide a tool that Contracting Parties and competent intergovernmental organizations can use to identify areas and features of the marine environment, both within and beyond national jurisdictions, that are important for conservation and the sustainable use of marine and coastal biodiversity (CBD 2012, X/29 para. 25). To assist in implementing this work, the CBD is sponsoring a series of regional workshops in conjunction with the FAO, regional seas conventions and action plans, and regional fisheries management organizations prior to COP 11 in 2012, with the primary objective of facilitating the description of EBSAs within and beyond national jurisdictions (CBD 2012, X/29 para. 36).

At the national level, COP 10 recommended that States further integrate climate-change-related aspects of marine and coastal biodiversity into national biodiversity strategies and action plans, national integrated marine and coastal management programmes, and the selection, design and management of marine and coastal protected areas (CBD 2012, X/29 para. 77). Finally, in its decision on marine and coastal biodiversity, COP 10 proposed convening an expert workshop with the UNFCCC on the role of marine and coastal biodiversity and ecosystems in adaptation to and mitigation of climate change impacts. The purpose would be to provide guidance for planning and implementing ecosystem-based approaches to climate change mitigation and adaptation and their integration in broader adaptation, mitigation and disaster risk reduction strategies (CBD 2012, X/29 para. 77). The focus on climate change impacts on marine and coastal biodiversity in the COP 10 decisions reflects an approach that seeks to incorporate climate change considerations into the traditional tools for ecosystem-based management of the marine environment, including the establishment of marine protected areas or areas in which special conservation measures are applied, as well as marine spatial planning.

4.2 Asia-Pacific Regional Law and Policy Frameworks

To avert some of the worst impacts of climate change and to mitigate its detrimental effects on the marine and coastal biodiversity of the Asia-Pacific, collaboration among the countries of the region, extra regional partners and global and regional organizations is needed at many levels. Some of this cooperation is already occurring. However, extension and innovative development will be required to reverse some of the adverse impacts of climate change on the environmental and economic security of the region. The following section will examine some of the regional initiatives to improve the resilience of marine and coastal biodiversity to climate change impacts.

4.2.1 Climate change initiatives in the Asia-Pacific region

In the Asia-Pacific, efforts are being taken to mitigate and adapt to the adverse impacts of climate change on the coastlines and marine biodiversity under the auspices of non-treaty-based regional environmental protection arrangements. Two such regional initiatives that have been taken in East Asia and the Asia-Pacific to protect the shared marine environment have strong climate change components. PEMSEA and the CTI reflect a common concern between East Asian States and some adjacent Pacific States for their shared marine environments in both the semi-enclosed seas of East Asia and the Pacific Ocean areas to the east of Japan and the Philippines. A group of 12 States and 15 non-State entities are partners in PEMSEA, which was established as a regional project of the Global Environment Facility (GEF) in 1994 with the initial aim of preventing and managing marine pollution in the East Asian seas (PEMSEA 2012a). PEMSEA's principal objective has developed into building interagency, inter-sectoral and intergovernmental partnerships for achieving the sustainable development of East Asian seas.

In November 2009, PEMSEA signed an agreement with the World

Bank to address the challenges posed by the growing populations of and continued rural migration to the coastal cities in East Asia, which are threatening the quality and sustainability of coastal life (PEMSEA 2012c). The partnership will complement the efforts of 11 East Asian States, the UN, the GEF and 19 regional partners in protecting mangroves and coral reefs, preventing overfishing, improving water quality, and creating greater preparedness for natural disasters and the effects of climate change. Information on local climate change adaptation strategies was exchanged between PEMSEA member States at an Experts' Forum on Climate Change Adaptation Strategies for Coasts and Oceans in the Philippines in early 2010 (PEMSEA 2012b). The initiative is still at an early stage and subject to resource and technical capacity limitations.

The CTI is another example of non-treaty-based maritime cooperation in the Asia-Pacific region, with a strong focus on climate change adaptation. The Coral Triangle is a region located along the equator at the confluence of the Western Pacific and Indian Oceans, which covers all or part of the exclusive economic zones of six countries: Indonesia, Malaysia, the Philippines, Papua New Guinea, the Solomon Islands and Timor L'Este. The Coral Triangle is regarded by scientists as one of the richest repositories of marine biodiversity on earth. It contains 76 per cent of all known coral species, 37 per cent of coral reef fish, and 33 per cent of the world's coral reefs. Further, it contains a wealth of mangrove forests and the spawning and juvenile growth areas for the world's largest tuna fishery (ARC 2008). Threats to the CTI region include overfishing, destructive fisheries practices, land-based sources of marine pollution and the ravages of climate change (CTI 2009). The CTI was proposed by Indonesia in 2007 as a multilateral partnership to protect the region's coastal and marine resources. Member States, Indonesia, the Philippines, Malaysia, Timor L'Este, Papua New Guinea and the Solomon Islands have committed to five overall goals over 10 years:

- the designation of priority seascapes
- the implementation of an ecosystem approach to managing fisheries and other marine resources
- the establishment of marine protected areas
- the development of strategies to adapt to climate change
- the protection of threatened species (CTI 2009).

The member States have committed to guiding principles including the recognition of the trans-boundary nature of important marine resources and the need to align their activities with existing international law instruments, such as the LOSC, CBD, regional fisheries management agreements and the UNFCC. Many of the regional and national actions under the CTI are contributing directly to climate change adaptation along the coasts and in the offshore maritime zones of the CTI region (CTI 2009).

In the Pacific, a key climate change policy initiative is the Pacific Adaptation to Climate Change (PACC) project. This project, funded by the GEF with the United Nations Development Programme (UNDP) as its implementing agency, and the SPREP as implementing partner, is designed to promote climate change adaptation as a key prerequisite to sustainable development in the Pacific Island countries and to enhance the capacity of the participating countries to adapt to climate change, including climate variability, in key development sectors (SPREP 2012). It is to be conducted over four years, from 2008 to 2012, and covers 13 Pacific Island countries. The aim is to build the Pacific countries' resilience to climate change by addressing the three key areas of food production and food security, coastal management, and water resource management. Adaptation projects are being implemented nationally.

Under the project, the Cook Islands, the Federated States of Micronesia, Samoa and Vanuatu are developing their coastal management capacity to adapt to climate change impacts. Climate change risks are being incorporated into relevant governance policies and strategies for achieving coastal development. At the sub-national level, pilot demonstration activities are being undertaken in the form of practical experience in the planning and implementation of response measures that reduce vulnerability to climate change impacts. The project will also foster regional collaboration on climate change adaptation. As with the PEMSEA climate change initiatives, the PACC is subject to significant resource and technical capacity constraints. As these regional initiatives evolve, the supplementary guidance on climate change adaptation strategies related to marine and coastal biodiversity being developed at the global level through the CBD and other international organizations should be channelled into pilot activities at the regional level.

4.3 National Law and Policy Frameworks – the Australian Example

The principal legislative authority for conserving marine and coastal biodiversity from threats, including climate change, is found in the 1999 EPBC. However, as climate change has emerged as one of the most prominent threats to marine and coastal biodiversity, more specific policy statements have been generated to define the threat and appropriate policy responses. Australia has attempted to address the causal links between climate change and declines in marine and coastal biodiversity through a hierarchy of high-level policy statements, coupled with the implementation of more concrete action plans. The National Strategy for the Conservation of Australia's Biodiversity (NSCABD), issued in 1996 and reviewed in 2001, is relevant because it set the broad parameters for the protection of Australia's biodiversity (NSCAB 2001, pp. 146–147). Some of the principles enunciated in the NSCABD relate specifically to the protection of marine and coastal biodiversity from the adverse effects of climate change and underpin the implementation of further action plans. These principles include:

- Biological diversity is best conserved *in situ*.
- It is vital to anticipate, prevent and attack at source the causes of significant reduction or loss of biodiversity.
- Lack of full knowledge should not be an excuse for postponing action to conserve biodiversity.
- The conservation of Australia's biodiversity is affected by international activities and it requires actions extending beyond Australia's national jurisdiction.
- Central to the conservation of Australia's biological diversity is the establishment of a comprehensive, representative and adequate system of ecologically viable protected areas, integrated with the sympathetic management of all other areas, including agricultural and other resource production systems (NSCAB 2001, pp. 146–147).

The NSCABD has now been supplemented by the Australian Biodiversity Conservation Strategy 2010–2020, which draws a more explicit link between the conservation of biodiversity and the impacts of climate change. A key objective under this strategy is 'to ensure our biodiversity is healthy, resilient to climate change and valued for its essential contribution to our existence' (Department of Sustainability 2010).

Beneath these overarching policy statements of Australia's biodiversity protection objectives, a *National Approach to Addressing Marine Biodiversity Decline* has been prepared by a Working Group convened by the Marine and Coastal Committee of the Natural Resource Management Ministerial Council. This document identifies the key threats to marine biodiversity from climate change and proposes priority actions for Federal and State Governments to implement in addressing these threats (Department of Sustainability 2008). Among the likely implications of climate change for the marine environment, the report highlights:

- loss or degradation of habitat, or changes in its distribution and density

- changes in ocean currents, upwellings and productivity
- displacement or distributional and abundance changes of marine species
- loss of synchronization between essential climate, weather and seasonal events affecting biota (such as a mismatch between phytoplankton blooms and zooplankton growth)
- lower ocean productivity and disrupted or changed food chains
- ocean acidification (changing the ability of calcium carbonate producing organisms to construct shells) (Department of Sustainability 2008).

Two of the key policy responses recommended by the report relate to climate change impacts on marine biodiversity and propose that Federal and State jurisdictions improve their understanding of the vulnerability of marine biodiversity to climate change, focusing on ecosystems and species that are at particular risk. Further, it is recommended that these jurisdictions develop regional climate adaptation policies and plans based on predictive modelling and integrate them into marine bioregional planning processes (Department of Sustainability 2008, p. 34). Priority actions proposed by the report include:

- identifying species and systems at particular risk from climate change (such as local endemics restricted to a small area of suitable habitat, like the spotted handfish) or exceptional ecosystems with unique evolutionary origins unlikely to be replicated in another area (e.g., Bathurst Harbour, south-west Tasmania)
- identifying processes threatened by climate change (e.g., tightly coupled processes that become decoupled due to changes in timing, chemical changes in the oceans caused by acidification, and coral bleaching caused by increased temperature maxima)
- developing regional climate models and scenario modelling to assess the potential effects of major regional climate change on marine activities (particularly fisheries and aquaculture) and biodiversity
- developing regional marine climate change adaptation plans that identify climate risks and vulnerabilities and also marine management scenarios and adaptations for marine industries and activities (fisheries, aquaculture and coastal development)
- integrating current knowledge of regional climate change risks and vulnerability into current large-scale bioregional planning and decision-making processes
- developing a national governance framework to assess and review the integration of current understanding of marine climate change

into marine management frameworks and directions (Department of Sustainability 2008, p. 34).

One of the key goals of the marine bioregional planning process currently being undertaken by the Federal Government in Australia is to improve the resilience of Australia's marine ecosystems so that they are better able to adapt to the impacts of climate change (Department of Sustainability 2011). More specific action plans are also being implemented for specific marine industries and areas. Under the National Climate Change Adaptation Framework, agreed upon by the Council of Australian Governments (COAG) in 2007, a five-year Climate Change Action Plan is underway to minimize the impact of climate change on the Great Barrier Reef through increasing its resilience. Additionally, a National Fisheries and Climate Change Action Plan has been endorsed by COAG and is in the process of implementation (Department of Sustainability 2008). The Australian policy response to the projected impacts of climate change on marine and coastal biodiversity has been intensive, but is still in the early stages of implementation. Future assessments will determine whether it has contributed to buffering Australia's abundant marine and coastal biodiversity from the worst effects of climate change.

The legal and policy infrastructure for conserving marine biodiversity from the adverse impacts of climate change is steadily growing at the global level, in regional organizations in the Asia-Pacific and through policy development at the national level, as in the Australian example. However, the most daunting challenge lies in effectively implementing the plethora of priority actions recommended to arrest the decline of coastal and marine biodiversity in the face of growing threats from climate change.

5. CONCLUSIONS

The challenges of marine and coastal management are not new, but they are difficult to address in the conventional framework of sectoral competition. The sectoral focus on biodiversity management has been on establishment of marine protected areas (Toropova et al. 2010) and the development of an ecosystem-based approach to the management of fisheries (FAO 2008). Recent attention on the complex policy issues of integration and coordination between sectors and jurisdictions is reflected in a growing number of publications on marine spatial planning and ecosystem approaches to marine management from the coast to the oceans (UNEP 2011; Ehler and Douvere 2009; Kidd et al., 2011; McLeod and Leslie 2009). Nevertheless, the legal and policy challenges associated with providing reasonably consistent frameworks for biodiversity management across marine areas within and beyond national jurisdiction remain substantial. While some regions have developed biodiversity conservation and management plans across areas within national jurisdiction through their regional sea programmes, gaps in coverage remain, and there is no comprehensive conservation and management system for marine biodiversity beyond national jurisdiction.

On-going efforts to address the impacts of climate change on marine biodiversity will benefit from the development of more integrated legal and policy frameworks for the conservation and management of biodiversity across national boundaries and beyond national jurisdictions. The Biodiversity Beyond Areas of National Jurisdiction process established by the UN General Assembly is likely to be a focal point for legal and policy development supported by scientific and technical advice from the CBD and the global marine science community. The burden of implementation in the Asia-Pacific and other regions will continue to rest with national jurisdictions working collaboratively to address the challenges of conserving and managing biodiversity in the face of climate change impacts across national boundaries and in proximate areas beyond national jurisdiction.

REFERENCES

ARC (2008), 'What is the Coral Triangle initiative?, *Coral Reef Studies*, accessed 26th July 2012 at: http://www.coralcoe.org.au/events/ctiworkshop/ctiinfo.html.

Birnie, Patricia, Alan Boyle and Catherine Redgwell (2009), *International Law and the Environment* (3rd edition), Oxford: Oxford University Press.

Boyle, Alan and Catherine Redgwell, *International Law and the Environment* (Third Edition) (Oxford: Oxford University Press, 2009).

CBD (2012), 'COP decision', accessed 26th July 2012 at: http://www.cbd.int/decision/cop/?is=12299.

Cheung, W.W.L., C. Close, K. Kearney, V. Lam, J. Sarmiento, R. Watson and D. Pauly (2009), 'Projections of global marine biodiversity impacts under climate change scenarios', *Fish and Fisheries*, **10**, 235–251.

Commonwealth of Australia (2007), *Climate Change in Australia. Technical Report*, accessed 12 March 2012 at: http://www.climatechangeinaustralia.gov.au/technical_report.php.

Connell, J.W., T.P. Hughes and C.C. Wallace (1997), 'A 30-year study of coral abundance, recruitment and disturbance at several scales in space and time', *Ecological Monographs*, **67** (4), 461–468.

CTI (2009), *CTI Regional Plan of Action*, accessed 26th July 2012 at: http://www.coralcoe.org.au/events/ctiworkshop/CTI%20Regional%20Plan%20of%20Action_June%2023%202009.pdf.

Department of Sustainability, Environment, Water, Population and Communities, (2008), *A National Approach to Addressing Marine Biodiversity Decline. Report to the Natural Resource Management Ministerial Council*, accessed 26th July 2012 at http://www.environment.gov.au/coasts/publications/marine-diversity-decline/index.html.

Department of Sustainability, Environment, Water, Population and Communities (2010), 'Australia's biodiversity conservation strategy 2010–2020', accessed 26th July 2012 at: http://www.environment.gov.au/biodiversity/strategy/index.html.

Department of Sustainability, Environment, Water, Population and Communities (2011), 'Fact sheet: marine bioregional planning – the process', accessed 26th July 2012 at: http://www.environment.gov.au/mbp.

Deser, C., A.S. Phillips and M.A. Alexander (2010), 'Twentieth century tropical sea surface temperature trends revisited', *Geophysics Research Letters*, **37**, L10701, doi: 10.1029/2010GL043321.

Douglas B.C. (1997), 'Global sea rise: a redetermination', *Surveys in Geophysics*, **18**, 279–292, doi: 10.1023/A:1006544227856.

Ehler, C. and F. Douvere (2009), *Marine Spatial Planning: A Step-by-step Approach Toward Ecosystem-based Management*, Paris: UNESCO, accessed 26th July 2012 at: http://unesdoc.unesco.org/images/0018/001865/186559e.pdf.

FAO (2008), *Human Dimensions of the Ecosystem Approach to Fisheries*, Rome: FAO.

Grubb, M., M. Koch, K. Thomson, A. Munson and F. Sullivan (1993), *The 'Earth Summit' Agreements: A Guide and Assessment*, London: Earthscan.

IUCN (1976), *An International Conference on Marine Parks and Reserves: Papers and Proceedings of an International Conference held in Tokyo, Japan, 12–14 May, 1975*, Morges, Switzerland: IUCN.

IUCN (1988), *17th Session of the General Assembly of IUCN and 17th Technical Meeting, San José, Costa Rica 1–10 February 1988 Proceedings*, Morges, Switzerland: IUCN.

IUCN/UNEP/WWF (1980) *World Conservation Strategy*, IUCN/UNEP/WWF.

Joyner, C.C. (1995), 'Biodiversity in the marine environment: resource implications for the Law of the Sea', *Vanderbilt Journal of Transnational Law*, **28**, 635.

Kenchington, R.A. and D. Crawford (1993), 'On the meaning of integration in Coastal Zone Management', *Ocean and Coastal Management*, **21** (2), 109–127.

Kidd, S.A., A. Plater and C. Frid (2011), *The Ecosystem Approach to Marine Planning and Management*, London, UK: Routledge.

Kimball, L.A. (1995), 'The biodiversity convention: how to make it work', *Vanderbilt Journal of Transnational Law*, **28**, 763.

McLeod, K. and H. Leslie (eds) (2009), *Ecosystem-Based Management for the Oceans*, Washington, DC: Island Press.

NSCAB (2001)*The National Strategy for the Conservation of Australia's Biodiversity*, Department of Environment and Heritage.

PEMSEA (2012a), 'About PEMSEA', accessed 26th July 2012 at: http://www.pemsea.org/about-pemsea.

PEMSEA (2012b), 'Climate change adaptation strategies for coasts and oceans', accessed 26th July 2012 at: http://www.pemsea.org/about-pemsea/pemsea-news/climate-change-adaptation-strategies-for-coasts-and-oceans/?searchterm=climate%20change.

PEMSEA (2012c), 'EAS Congress 2009', accessed 26th July 2012 at: http://www.pemsea.org/eascongress/.

SPREP (2012), 'Pacific adaptation to climate change', accessed 26th July 2012 at: http://www.sprep.org/climate_change/PACC/Country.htm.

Toropova, C., I. Meliane, D. Laffoley, E. Matthews and M. Spalding (eds) (2010), *Global Ocean Protection: Present Status and Future Possibilities*. Brest, France: Agence des aires marines protégées.

UNEP (2011), *Taking Steps Toward Marine and Coastal Ecosystem-Based Management: An Introductory Guide*, UNEP Regional Seas Reports and Studies No 189, Nairobi: UNEP.

Webb, A. (2010), 'Atoll shoreline response to sea level rise over the last 50 years (Pacific Regional Island Shoreline Monitoring System)', in *Proceedings of the International Seminar on Islands and Oceans*, Tokyo: Ocean Policy Research Foundation, pp. 63–65.

World Bank (2006), *Scaling Up Marine Management: The Role of Marine Protected Areas* (August), The International Bank for Reconstruction and Development/World Bank Report no. 36635-GLB.

4. Climate change and marine living resources

Pia C. Winberg, Ana Rubio and Lisa Kirkendale

1. INTRODUCTION

Marine living resources encompass the full suite of marine biodiversity that provides for ecological services, such as bio-geochemical cycling of nutrients, commercial operations such as coral reef tourism, and primary production from fisheries and aquaculture. This chapter will focus on the last of these, marine living resources for food production, and infer the practical challenges that climate change will place on these resources across subsistence, artisanal and global fisheries and aquaculture operations and in different socioeconomic conditions in the Asia Pacific.

Asia Pacific fisheries and aquaculture are as diverse in species, scale, technology and production as they are culturally and climatically. Despite the diversity of natural marine resources, land masses, cultures and socioeconomic conditions, the Asia Pacific collectively includes the three largest seafood producing regions globally: the Northwest Pacific, the Southeast Pacific and the Western Central Pacific (FAO 2010b). Millions of people in Asia rely on seafood production for subsistence, as well as for local and export markets. Outstanding growth in the marine aquaculture sector, particularly in China, has transferred the balance of seafood production towards aquaculture in many Asian countries (see Figure 4.1). Despite this, China and five other Asian countries remain in the top 10 list of nations with the largest wild fisheries in the world (FAO 2008a). China stands out globally in terms of seafood production and consumption – unsurprisingly, considering it has the largest human population with a high per capita consumption of seafood. China is shifting from becoming a net producer to a net importer of seafood, placing enormous pressure on regional and global seafood demand as a result.

In contrast to the Asian nations, the Pacific is still heavily dominated by wild fisheries with less adoption of aquaculture technologies. Smaller,

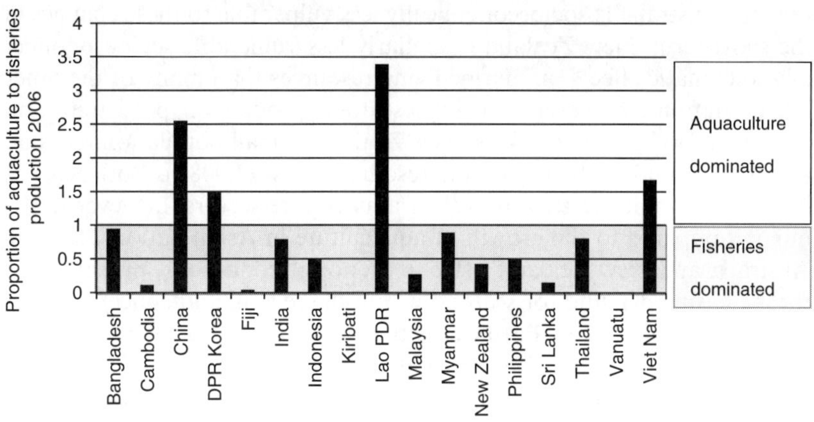

Source: Adapted from Lymer et al (2008).

Figure 4.1 The proportion of aquaculture versus wild capture production of seafood across 17 Asia Pacific nations

isolated populations in the Pacific rely on artisanal and part-time fisheries for food production, as well as having economic reliance on external access fees from an international fleet of large vessels (FAO 2008a; Kompas 2008). With the exception of Papua New Guinea, most Pacific Island nations have small landmasses, and food security from the oceans is critical. Even the island nations with the lowest fish consumption far exceed the average global seafood consumption of 18 kg per person each year; Pacific Island seafood consumption ranges from 30–146 kg per person per year. Some regions may be resilient to or even benefit from climate change scenarios with maintained or improved oceanic productivity; while others may face reduced production and serious threats to their economies and food security. With little land or other industries to rely on for income and food production opportunities (Kronen et al. 2010), these nations must establish diverse strategies to maintain food and economic security from the oceans, including setting sustainable human population targets, diversifying production technologies (e.g., some transition to aquaculture practices and other income opportunities), addressing resource and income management structures, and monitoring and modelling ecological shifts in response to climatic changes.

South eastern Australia is already experiencing some of the most rapid and measurable changes to living marine resources in response to climate change (Lough 2009). However, as Australia is a low consumer and producer of seafood compared to its Asian and Pacific neighbours (ABARE 2009), with significant terrestrial food production resources to rely on, Australia is socioeconomically less vulnerable to these changes in the short term. New Zealand is similarly less vulnerable socioeconomically to climatic effects on marine living resources than many of the other Pacific nations, however with high seafood production per capita and strong seafood export markets, New Zealand's small population relies on access to abundant living marine resources (FAO 2008b). Both nations have seen significant growth in their aquaculture sectors, however this is minor compared to the growth of aquaculture in Asian nations. Despite Australia and New Zealand's strong economic conditions, preparedness for increasing demand for seafood within the region, and concern for the welfare of neighbouring Pacific countries, is a social responsibility and an issue of regional security for these nations.

The advent of climate change adds another layer of complexity to the increasing demand for seafood and the historic challenges of achieving sustainable fisheries and establishing new and viable aquaculture operations. More variable and erratic production is anticipated in the wild fisheries sector due to climatic change and this is already being felt (Hobday and Poloczanska 2010). Such changes include altered marine food chains, shifting productivity, changes to breeding patterns of commercial species, and changes in species distribution. However there is also evidence of new opportunities as a result of climatic change, particularly in unison with the growing aquaculture sector and a need to restructure seafood industries to work within sustainable limits (Johnson and Welch 2010).

This chapter seeks to synthesize the complex but demonstrable climate change effects on marine living resources relevant to seafood production from peer-reviewed literature and intergovernmental assessments. The purpose is to translate substantiated climate change effects into a suite of tangible impacts on aquaculture and fisheries sectors (see Table 4.1). The chapter is structured into 11 specific aspects of climate change impacts on fisheries and aquaculture, including production and yield, disease management and genetic considerations. A range of response options to each of the types of impacts are considered, including the policy implication of these options.

2. IMPACTS AND IMPLICATIONS FOR FISHERIES AND AQUACULTURE

2.1 Changes to Production and Yield

It is well established that oceanic water masses are distinct in terms of productivity. It is estimated that 90 per cent of productivity in the oceans

Table 4.1 The perceived or established relationship between the physical, chemical and biological effects of climate change on the aquaculture and fishery industries

Climate change effects	Physical						Chemical					Ecological/Biological						
Direct industry impacts	Air and sea temperature	Hydro-dynamics	Rainfall	Severe weather	Sea-level rise	Sedimentation	Salinity	Acidification	Nutrient supply	Turbidity	Oxygen	Primary production	Harmful algal blooms	Species distribution	Genetic resources	Introduced species	Recruitment/seed supply	Disease
Production and yield																		
Type of yield																		
Growth and mortality																		
Pests and fouling																		
Seed supply/recruitment																		
Species shifts																		
Genetic diversity																		
Coastal assets																		
Lack of feed																		
Human health																		
Harvest and market regularity																		

Note: Light to dark shades indicate qualitative and relative strengths of interaction from light (low) to dark (high).

is fixed by marine phytoplankton (Behrenfeld et al. 2006), and that the 100M tonnes of carbon fixed each day is a pre-cursor for the growth of the biomass of fish and aquaculture stocks. Consequently, primary production and food-chain productivity are spatially correlated (see Figure 4.2) and the pattern of primary production is one of the most critical components to the viability of the seafood industry (Chassot et al. 2010; Klyashtorin 2001). For example, in most years, prevailing La Niña conditions deliver high production to the western Pacific islands, including Papua New Guinea, while during El Niño conditions, the eastern nations around Kiribati benefit from the tuna stocks that respond to the

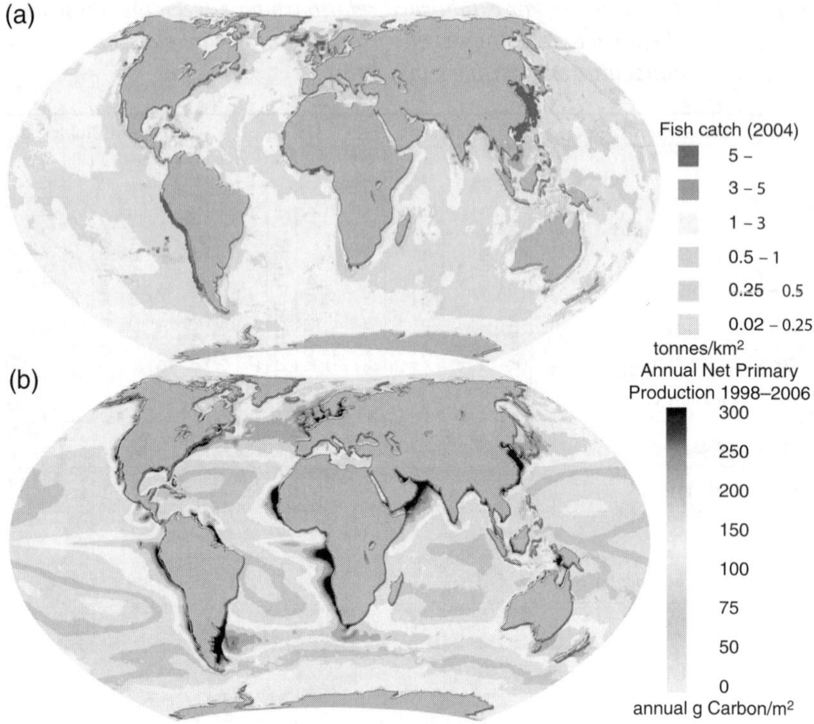

Notes: Map (a) shows annual catch (tonnes per km^2) for the world's oceans, while Map (b) illustrates regional primary productivity. Primary production in the oceans was recorded from 1998 to 2006 by VGMP/SeaWIFS data and was interpreted by Oregon State University. Highly productive areas are reflective of oceanic upwelling along geological barriers.

Source: Figure provided by Hugo Ahlenius, UNEP/GRID-Arendal.

Figure 4.2 Global fish catch hot spots correlated to areas of high primary productivity and ocean currents that transport this production

spatially shifting food chains (Bell et al. 2011). Regions of lower productivity are confronted with a higher effort and therefore a higher cost of seafood production per tonne, and so have not developed as critical industries to the local national economies; nevertheless in smaller nations relatively low production waters still make an important contribution to regional economies, socioeconomic stability and local food production (Wongsanga and Vichitlekarn 2010).

The spatial patterns and timing of primary production in the oceans are predicted to change significantly with the changing climate, in large part in relation to changes in nutrient delivery and 'the other global change' – ocean acidification (Beardal and Raven 2004; Doney et al. 2009). Similar impacts of such primary production shifts have been evident since the 1980s (Chassot et al. 2010), with changes to reef fish in the Pacific and French Islands suggested to be a response to a period of declining primary productivity over decades (DeMartini et al., 1996). Declines in primary production in the oceans will have negative consequences for seafood production in some regions (Feely et al., 2004), while some regions will shift towards increased productivity (Beardal and Raven 2004; Levitan et al. 2007), with positive outcomes for existing and new seafood opportunities. In either case, the contention that shifts in primary production due to climate change will have consequences for the living marine resources that we rely on for food is supported (Beaugrand et al. 2010; Hallegraeff 2010), implying that monitoring of trends in productivity will be important to pre-empt catastrophes in the Asia Pacific region.

2.1.1 Role of policy in response to reductions in production

Currently, up to 22 nations within the Pacific Micronesian, Melanesian and Polynesian regions share spatially fluctuating purse seine fisheries resources and receive access fees from international fleets that operate in response to areas of high productivity. These fleets catch about 2 million tonnes of pelagic tuna for export annually, contributing about US$200 million to the local economies, including government structures (Bell et al. 2011), and US$1 billion to international market economies. However, there is ongoing debate as to whether this form of management is in the best interests of sustainable fisheries and the local communities as it is driven heavily by demand from and profits for the west.

In addition, many Pacific Island artisanal fisheries rely on the productivity of coastal food chains to provide for protein production for local communities. These fisheries are often overlooked and undervalued due to their disparate nature and the difficulty in quantifying effort and production volume. Each of the diverse jurisdictions and scales of seafood industries have specific and unique interests and vulnerabilities in relation to climate change (Kronen et al. 2010; Wongsanga and Vichitlekarn 2010). Consequently, governance and policy, including initiatives to reduce the overcapacity of the fishing sector, does and will play a central role in the region's ability to adapt to climate change. There is consistent reference in the scientific and policy literature to the failure of regional fisheries management organizations (RFMOs) to control the effort of wild fisheries in the high seas and in proximity to these small nations (Cullis-Suzuki and Pauli 2010). However, the onset of climate change impacts may deliver an opportunity to develop new approaches and solutions to fisheries management and/or reinvigorate community-based marine resource management to promote firmer control over international fisheries operations in local waters (APFIC Asia Pacific Fishery Commission 2010; Johannes 2002).

The ability of fisheries to adapt to changing ocean productivity will depend on their capacity to predict stock recruitment, spatial patterns of productivity for existing and new fishing grounds, the distance to productive fishing and aquaculture areas and the capacity to travel and work in these areas (Hobday and Poloczanska 2010). To better inform current and future management and policy initiatives of the shifts in productivity and changing trends, ongoing productivity monitoring networks are essential to maximize foresight and predictive power, including the monitoring of trends in oceanic primary production and its correlation with seafood yields. Such networks and information dissemination need to involve collaboration between government, industry and science organizations in the interests of industry productivity and socioeconomic stability.

The transition to aquaculture from wild fisheries is also one of the responses to shifts in oceanic primary production that is ongoing, and it is predicted to continue in response to the changing climate. Fed aquaculture species have the benefit of being less reliant on local oceanic productivity regimes and therefore might be suitable in areas of fluctuating wild stocks. However, analogous to sustainable fishing effort, the carrying capacity of aquaculture operations in lease areas need to balance input and removal of nutrients and biomass must be determined and be continually monitored and re-modelled under the changing climate scenarios.

2.2 Changes to Type of Production or Regime Shifts

The effects of climate change on ocean productivity described above and placed in the context of our knowledge of oceanic trophic food-chains, strongly indicates that climate change effects may further exacerbate changes to the targeted seafood species from fishing down the food chain to smaller species (Mollmann et al. 2008; Pauly et al. 1998; Shackell et al. 2010). Such impacts are also known as regime shifts, and in the Asia Pacific as in other oceans, an analogous situation is emerging with abundant planktonic feeders, especially jellyfish, on the increase (Atrill et al. 2007; Uye 2008; Wianans and Purcell 2010). Reflecting on potential changes to primary production, as discussed above, local enhancement of planktonic feeders can result in a shift towards more abundant smaller fish species in catch landings (Uye 2008), leading to closure of fisheries, and have either positive or negative consequences for aquaculture (Richardson et al. 2009) – especially for shellfish and other low food chain species.

2.2.1 Role of policy in responses to changes in type of production

While governance and policy initiatives aimed at monitoring fish stocks and catch size trends are already in operation, these should be extended to consider shifts in other trophic levels of ocean biomass. As much as 30 per cent of fisheries landings in the Asia Pacific region are classified as 'unknown' species (APFIC Asia Pacific Fishery Commission 2010), and shifts in the type of production will therefore be difficult to detect without improved monitoring.

The promotion of more sustainable fishing at lower food-chain levels is one approach that can serve to improve both fisheries management and the capacity to adapt to a higher abundance of smaller and low food-chain species. In part this is already happening due to overexploited high level food-chain species, but also in many Asian countries low food-chain species are well accepted and dominate fisheries and aquaculture. In developing nations, very small fish can be a vital source of complete nutrition (Larsen et al. 2000) and are considered a more sustainable supply of seafood, as the biomass of seafood is larger and more resilient at lower levels of the food chain. However, western nations will require education and marketing campaigns to promote the consumption of smaller and lower trophic-level species.

2.3 Changes to Growth and Mortality

Mortality is a usually a manageable and natural event in populations of any species, and is a result of an interaction between an organism, pathogens and environmental factors. In contrast large-scale disease outbreaks or mass mortality events have costly consequences for industries that rely on marine living resources, and can also threaten food security. With the onset of climate change, changes in disease and/or mortality events can be expected owing to shifts in environmental stressors (Arnold et al. 2009; Portner and Knust 2007). Environmental stressors can operate on both the condition of the host organism as well as the abundance and viability of pathogens in the environment (Battaglene et al. 2008). Therefore, the interaction between a shifting environment, pathogen abundance and host density, especially in the case of aquaculture, suggests that disease is a key threat to seafood production in certain areas (Anestis et al. 2010; Krkošek 2010). For example, the increasing temperature on the east coast of Tasmania appears to be impacting on salmonid growth and health in aquaculture (Battaglene et al. 2008), and there is increasing disease frequency in species of the Asia Pacific including corals (Sutherland et al. 2004), shellfish (Anestis et al. 2010; Tomaru et al. 2001) and pelagic fish (Crockford 2007). The bivalve (e.g., mussels, scallops, oysters and clams) aquaculture industry in the Asia Pacific is thought to be under particular threat from the environmental effects of climate change, as this form of aquaculture is reliant not only on optimal growth conditions in the environment, but also on the production of a natural food supply for optimal health (Canu et al. 2010), both of which can shift in response to climate change and render bivalves vulnerable to disease.

The introduction or proliferation of pathogenic species in oceanic water masses can occur either through anthropogenic-translocation activities across the oceans, or through changes in oceanic currents and physical and chemical gradients (e.g., temperature). In addition, fish species as hosts have distinct parasitic and pathogen assemblages, which have co-evolved (Niklitschek et al. 2010), and spatial shifts in host species populations to new regions (discussed further below) will also include shifts of pathogens and expose existing populations to new diseases.

However, pathogens are not the only cause of mortality in fish species, and biotoxins from harmful algal blooms (HABs) are also important. In a review of HAB impacts in the Asia Pacific in the last century (Corrales and Maclean 1995), 57 per cent of HABs were found to be the cause of mass fish or shrimp kills, with huge economic consequences. Therefore, the likelihood of an area being affected by HABs will need to be considered in identifying viable locations for aquaculture development. It is thought that currently suitable aquaculture areas – for example, for shellfish – are at risk of becoming less suitable due to climate change (Canu et al. 2010), while currently less suitable locations might improve in the future.

2.3.1 Role of policy in response to reduced growth and mortality

Based on the current knowledge of the interacting factors of environment, pathogens and disease and mortality, policy and legislation needs to target the integration of information from the monitoring and management of environmental parameters with biosecurity monitoring of fish stocks and aquaculture species. Continued improvement or the establishment of good management systems for biosecurity and the requirements of good animal husbandry and management in aquaculture will continue to be vital to the security of these industries.

In terms of the environment, the continuation and expansion of rigorous monitoring systems and the translation of the collected data into long- and short-term trends will be critical in informing decision-making processes. Once links can start to be made between the environmental fluctuations and stock health, disease and mortality, this will contribute to the capacity to pre-empt and cope with or adapt to change. Coping with or adapting to change can include management options for both aquaculture and fisheries such as the flexibility of lease or fishing areas for short- or long-term response to unfavourable climatic events.

Other important policy initiatives to deal with change and stress of marine living resources includes establishing suitable locations and carrying capacities of aquaculture. This is an urgent priority that is necessary for the long-term viability of existing and new developments in aquaculture. The Chilean salmon aquaculture industry has already experienced a dramatic mass mortality linked in part to the absence of policy and regulation about the carrying capacity of aquaculture. There were consequently severe effects for the local environment and socioeconomic conditions in coastal communities where aquaculture had become a key industry. Governance systems in Chile have recognized, through experience, the importance of health and environmental regulations in aquaculture and of establishing the limits to the carrying capacity of the industry (Alvial 2010). This and similar experiences throughout Asia during the exponential growth phase of aquaculture in the latter 1900s serves to demonstrate the importance of health and environmental governance-led initiatives throughout the Asia Pacific, where aquaculture development is still growing and will be very important to the local economies, food security and health.

In terms of managing pathogens, high standards of consistent biosecurity for all transnational, maritime industries are critical, and policies will need to continue to improve industry practices through stringent but fair environmental and biosecurity management systems. This will include the import of biological material for aquaculture, including stock and feeds, and, potentially, the designation of disease free zones (Subasinghe et al. 2004). Biosecurity policy and regulation extends beyond marine living resource industries. Disease introduction through transport, and maritime infrastructure can be controlled to a degree through disease surveillance programmes (Subasinghe et al. 2004), quarantine systems and transport management practices (e.g., ballast water management and hull and gear cleaning), which will become more important as the climate changes. Response management systems to disease outbreaks are also important and include requirements such as rapid and well-planned investigations, pathology responses, temporary bans on the import/export of produce and the translocation of gear to contain disease outbreaks.

For management of animal health issues, medication should be considered a last resort to good management practices (Rigos et al. 2010). Instead, a preventative approach through vaccination programmes and increased resistance to disease through selective breeding in aquaculture are preferred options; however, good animal husbandry and sustainable stocking densities remain of the highest priority.

Restocking programmes for wild stocks are undertaken for a few fisheries and selective breeding against disease could be applied to fisheries as well as in aquaculture (Parker et al. 2010). However there are limitations and risks that need to be considered in breeding programmes (Camara et al. 2008), and breeding for restocking programmes of wild fisheries is regarded as a last resort for effective fisheries management, which, in most instances, does not meet expectations (Bell et al. 2006). Generally, well managed and genetically diverse fish stocks are more resilient to disease and climatic fluctuations.

2.4 Impacts through the Introduction of Pests and Fouling

The introduction of non-indigenous marine species (NIMS) to a bioregion can occur without apparent impacts to the ecosystem or marine living resources, and, in many cases, these are cryptic. However, a large number of introduced marine species have had a huge impact on both ecosystems and biological diversity, with economic consequences for fishing and aquaculture industries (Bax et al. 2003). In a few rare instances, non-native marine species have become a valuable marine living resource, as seen by the example of the use of *Undaria pinnatifida* in Australia for food and nutraceuticals (Aquenal Pty Ltd 2008).

Vectors for introduced marine species include ship hulls, ballast water, recreational vessel hulls (Goldstien et al. 2010) and sea chests (Coutts and Dodgshun 2007). In addition, artificial hard surface infrastructure around ports and aquaculture facilities provides a suite of hard clean surfaces in the marine environment that act as a settlement vector for introduced marine species (Bax et al. 2003). There are also aquaculture industries in which the translocation of species or biological material is a common practice, with evidence for invasive species impacts (Haupt et al. 2010). While climate stability might have historically prevented the establishment of warmer water species that have been translocated, the onset of climate change and potentially more favourable conditions could provide for an increase in the successful establishment of pests, and local expansion post-invasion (Goldstien et al. 2010). In addition, climate change itself is a cause of some new species introductions as environmental regimes shift and take species with them (Banks et al. 2010).

2.4.1 Role of policy in response to pests and fouling

Responses to introduced marine species and pests include increased efforts in biosecurity, as outlined above for disease. For example, staging of ports of call for international vessels, standardized and effective hull and gear treatment policies, and strictly administered and enforced translocation protocols (Bax et al. 2003) will all contribute to reducing the risk of introducing of new species. In addition, marine pest surveillance, education, response and control plans must be further developed and practised for all pest outbreaks, native and introduced alike.

In some instances, new species introductions will provide opportunities for exploitation as marine living resources, and can be considered in policy terms as an adaptive response to climate change and the acceptable use of introduced marine species. However, such policy can be controversial, and an approach that recognizes the risks and management of such an approach independently for different species is required.

2.5 Changes to Seed Supply and Recruitment

It is expected that shifts in marine environmental conditions will affect the reproductive timing, output and larval success of important marine living resources. The importance of environmental cues for reproductive behaviour is well established and such cues are regularly applied to trigger the spawning of fish and invertebrates under hatchery conditions. In cooler latitudes of the oceans, positive recruitment responses to increased ocean temperature have been indicated for shellfish (Shepard et al. 2010). However, negative recruitment effects are also predicted. Coral reef fish, for example, have reduced reproductive behaviour and lower viable sperm production in elevated temperatures (Donelson et al. 2010). Some fish species exhibit temperature dependent sex-determination mechanisms, and a shift of only 1–2°C can significantly alter the sex ratio from 1:1, to up to 3:1 (Ospina-Ålvarez and Piferrer 2008), with significant consequences for reproductive success.

Apart from reproductive success, it is thought that the larvae of many organisms will be the most sensitive life stage to ocean acidification and thereby affect recruitment success (Fabry et al. 2008; Feely et al. 2004; Orr et al. 2005; Riebesell et al. 2000). For example, simulation studies have demonstrated negative effects of reduced pH on the larval success of commercial species such as oysters (Watson et al. 2009), abalone and sea urchins; an effect that might be enhanced in warmer regions of the oceans (Byrne et al. 2010). Starvation at the critical early-life stages of species has also been recorded as an impact related to climate change, through, for example, the mismatching of the timing of food supply (zooplankton) and the larval development stages of fish reliant on zooplankton (Yang and Rudolf 2010).

2.5.1 Role of policy in response to changes in seed supply and recruitment

Changes to the recruitment of marine living resources will exacerbate any effects of overfishing and could lead to stock depletion (Dudgeon et al. 2009). Therefore, the development of a better understanding of population

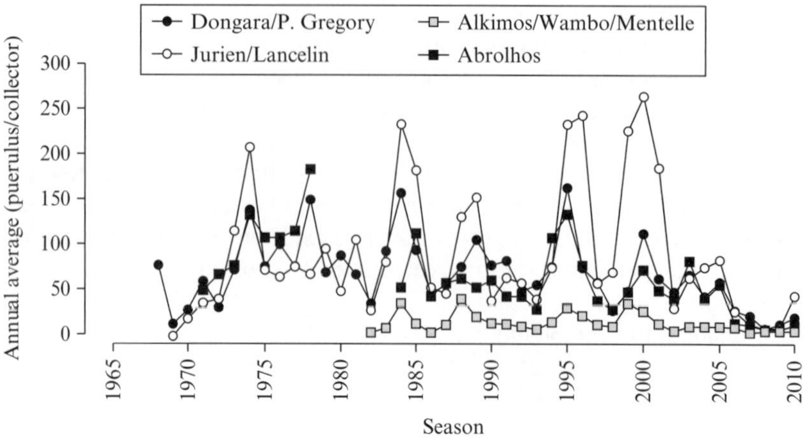

Source: Western Australia Department of Fisheries (2010).

Figure 4.3 Juvenile puerulus settlement of rock lobster since 1966 at four collection sites in Western Australia

structures from juveniles to adult stocks through fisheries monitoring programmes is a priority response to climate change. This is in line with the understanding that improved fisheries monitoring and management is a key policy approach that will provide for increased resilience and adaptation to climate change (World Bank, 2009). Some consider stock recruitment to be the most variable parameter in fish-stock assessment, and claim that it is very difficult to monitor (Fossa et al. 2008). However, for certain fisheries resources, monitoring has been shown to be effective in predicting the future biomass against which to set sustainable levels of fishing effort. For example, one of the most successfully managed fisheries globally – the Western Australian lobster industry – has invested in intensive monitoring and temporal assessment of recruitment to the population (Pecl et al. 2009; WA Fisheries 2010). Recent trends in this data have indicated that the recruitment to two separate populations of rock lobsters, in Tasmania and Western Australia, have declined in recent years, despite well-managed adult stocks. This early evidence of an unexplained change, which could be a result of changing ocean conditions, provides an opportunity for a proactive response. Scientists are now attempting to establish whether lobster juveniles are simply recruiting to new areas in response to changing currents, or whether there are other causes linked to climate change. Thus, this example clearly demonstrates the benefits of preparedness through monitoring programmes, which provide the adaptive capacity to change (see Figure 4.3).

Of further importance to seed supply and recruitment for many commercial fisheries and aquaculture industries is the integrity of the coastal and estuarine habitats that are critical to the early-life stages of many species. Poor coastal management and loss of critical habitats such as salt marsh, seagrass and mangrove are already acknowledged as major anthropogenic stressors on recruitment (Done and Reichelt 1998). Therefore, any further stress placed on these recruitment habitats through climate change may have significant adverse impacts for a number of species. Improved coastal and catchment management is therefore required for more resilient nursery and recruitment habitats.

Restocking and stock enhancement of wild fisheries has proven itself in a few circumstances; in particular for shellfish industries such as the scallop fishing sector in Japan (Uki 2006). For species like scallops and bêche-de-mer (sea cucumbers) –considered of high value, with the potential for expansion through aquaculture and of substantial importance to large socioeconomic sectors in the Asia Pacific – the application of breeding and restocking programmes may be viable. However, there are considerable costs associated with hatchery production (Lovatelli 2004), meaning that restocking and stock enhancement programmes through hatcheries will be useful for fisheries in only a limited number of circumstances. In addition there is the concern in relation to a limited understanding about the potential effects from the release of genetically altered strains (Bell et al. 2006). Restocking cannot be considered a long-term solution in instances in which climate change or environmental impact is the cause of reduced seed supply, and where environmental conditions are not improved (e.g., through catchment management of habitat restoration). In contrast, selective breeding and hatchery development is central to the development of viable aquaculture production, as evidenced by the global salmon aquaculture industry. Aquaculture will increasingly become reliant on hatchery-produced juveniles, but some wild recruitment industries remain and these will contribute an important source of genetic diversity and resilience to domesticated stock.

2.6 Shifting Species Ranges

Marine ecosystems exhibit clear geographical patterns of biodiversity, discrete ecosystems and spatially restricted populations due to environmental gradients, physical and chemical boundaries and limitations to species dispersal (Lessios and Robertson 2006). Distinct biological and fishing regions are recognized throughout the Asia Pacific (Dudgeon et al, 2009) and, despite cultural seafood preferences, the contrasting diversity of fish species available at local fish markets is evidence of this. It is thought

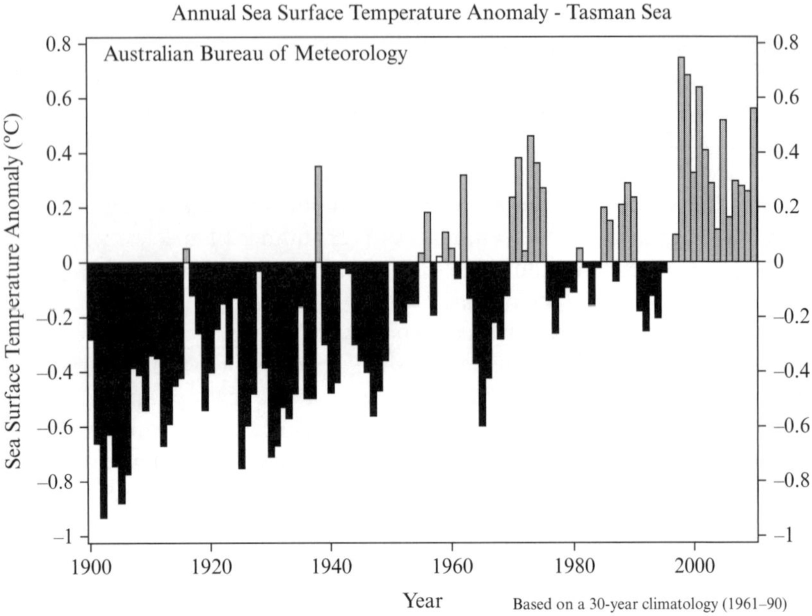

Source: Figure provided by the Australian Bureau of Meteorology.

Figure 4.4 The annual sea-surface temperature anomalies in the Tasman Sea during the past 100 years

that climate change effects will create, shift and/or break down existing species and population barriers and, in synergy with human induced translocation of organisms, substantial shifts in species distribution and range sizes are to be anticipated (Ling et al. 2008; NCCARF 2009).

The south eastern Australia region is a climate change hotspot, with well-documented changes having already occurred over the past 70 years. One example of this is the southward penetration of the East Australian Current by about 350 kilometres and a temperature rise of almost 2°C (see Source: Figure provided by the Australian Bureau of Meteorology. *Figure 4.4*). Concurrent with this change is a geographical shift southwards of 30 per cent of the inshore fish families occurring in south east Australia (Last et al. 2011).

Species range shifts can include large-scale redistribution of global fisheries as well as local fisheries catch potential. At a large scale, the generality appears to be population shifts towards the poles as a consequence of observed trends in ocean warming (FAO 2009). Cold-water species will be under stress, whereas warm-water species are thriving and expanding their distributions with the arrival of many non-indigenous species of warmer affinity in temperate and sub-polar regions (Hiddink and Ter Hofstede 2008). Areas of currently warmer conditions are expected to decrease in diversity and abundance, potentially losing up to 40 per cent of catches in some predictions (Cheung et al. 2010).

2.6.1 Role of policy in response to range shifts

The recognition of species range shifts as a consequence of climate change needs to be recognized in fisheries management, monitoring and response strategies, and considered concurrently with impacts from fishing effort. Further, evidence for species-specific range shifts is required, so as not to be confused with the impacts of fishing. Some of the smaller, artisanal and subsistence fisheries are spatially managed through historic knowledge of species range extents and community-based marine resource management (Johannes 2002). The contribution of such fisheries to regional countries in the Asia Pacific should not be underestimated (Kronen et al. 2010) and changes to species ranges will present challenges to current management practices. An immediate need is the formal recognition of spatial management systems for smaller-scale fisheries and how these might be impacted by range shifts. Existing spatial management structures may become less effective or require some flexibility and co-management solutions across jurisdictions and geographical boundaries.

The logistical and infrastructural implications of range shifts relate to the location of fishing ports and travel distances for vessels. Larger vessels may be more adaptable to geographical shifts due to their ability to travel to different locations or further afield. However rising fuel costs, which in some cases already comprise over 30 per cent of fishing costs, will have an impact on this adaptation potential (ABARE 2007), especially for smaller vessels and local fisheries (Kompas 2008). Relocation or adaptation to new species opportunities might be required to adapt to such potential changes.

The diversification, or increased types, of targeted species in both fisheries and aquaculture will be one option for adaptation, but this will require flexible governance and policy. For example, the southward encroachment of the northern sea urchin and rock lobster populations in Australia might result in a shift to harvesting of these species in new range extensions (Pecl et al. 2009) and introduce opportunities for new industries, including aquaculture. However, caution must be applied in exploiting such 'new' resources, as the genetic diversity of species undergoing range expansion can be low and they can be less resilient to other impacts (Lind et al. 2007). Indeed, the variable nature of any new assemblages requires conservative approaches to exploitation.

2.7 Impacts to and Implications of Genetic Diversity in a Changing Climate

Low genetic diversity of populations can compromise the capacity of a species to adapt to climate change. Due to other impacts or changes in a stable environment, the diversity of some populations is already compromised or low (Somero 2010). Thus, the impacts of climate change can contribute to further reduction of the genetic diversity of populations. For example, populations can be reduced through mass starvation events from climate effects (e.g., the case of the South American Fur Seal, see De Oliveira et al. 2009), or become isolated by the introduction of environmental barriers to gene flows (such as in the case of the Pacific Sierra, see Lopez et al. 2010). Some species have the genetic and physiological capacity to adapt to thermal changes, while others have lost that capacity through evolution in a thermally stable environment. The species considered to be most at risk include deeper water stenotherms from colder climates or eurytherms at the limits of their high thermal range (Somero 2010).

Climate change effects on genetic resources interact strongly with fisheries management. Some populations may be more or less resilient to fishing pressure than others, depending on the status of the genetic structure within populations; for example, deep water lobster appear to be more susceptible to fishing pressure than some of their shallow water counterparts (Palero et al. 2010). Genetic techniques can now provide powerful tools that can determine population status, vulnerability and boundaries to be applied to fisheries management (Ovenden et al. 2009).

Fisheries and aquaculture can also contribute to the selection of genetic traits in both bred and wild populations. Within aquaculture, genetic selection has benefits for domesticating breeds with desirable traits for markets, disease resistance, fast growth, or even enhanced resistance to climate change effects (Lang et al. 2009; Tester and Langridge 2010; Yan et al. 2007). However, default genetic selection through ad hoc aquaculture and fishing practices can render populations vulnerable. For this reason, a better understanding of genetic diversity is essential to fill this critical gap in current knowledge.

2.7.1 Role of policy in response to genetic diversity

Considering the potential consequence of fisheries and climate-induced genetic impacts, an improved understanding of the genetic structures of marine living resources is important to identify populations that are vulnerable to the effects of climate change. The current boom in cost-effective genetic technology and tools has provided powerful tools and delivered new information that is of great value to fisheries management and the identification of vulnerable populations. Genetic tools are becoming increasingly cost-effective and will contribute significantly to a greater capacity to meet the challenges of climate change.

2.8 Impacts to Coastal Assets

Land-based infrastructure for aquaculture and fisheries is considered particularly at risk from climate change impacts. This is because such infrastructure is necessarily placed along the coastal fringe (Simms and Woodroffe 2008) and is thus exposed to changes in sea level in conjunction with the increased frequency of extreme weather (Church et al. 2006a, 2006b). Of particular concern is that the land available in coastal areas is already limited and alternative sites are difficult to access. In developing island nations of limited land space, the potential impact on coastal assets and the limited options for land retreat of communities is a potentially major challenge (see Bell et al. 2011, Ch. 3).

In addition to infrastructure, natural coastal assets are very important for both fisheries and aquaculture, for example local fishing and nursery habitats (wetlands, salt-marshes, creeks and mangroves), and estuarine-based industry lease areas that are suitable for aquaculture gear and that provide a clean water supply. Many of these natural resources already suffer from intensive development along the coast and are predicted to be further impacted as a result of climate change. Specifically, increased storm events and sea level rise are predicted to have adverse impacts on sedimentation, nutrient input, pesticides and other pollutant discharge to local waterways, as well as affecting the extent of transition habitats for marine living resources (e.g., mangroves, salt-marshes and seagrass beds).

In addition to the expected physical impacts on coastal assets, sea-level rise and reduced river flows are already believed to be causing increasing saltwater intrusion into many estuaries. For example, aquaculture in the Mekong delta produces about 1 million tonnes of catfish per year, valued at $1 billion. The increasing salinity in this river is a threat to over 150 000 livelihood opportunities, in particular for women (PaCFA 2009). In Australia, the oyster industry exists predominantly at the mouth of estuarine systems, but relies heavily on regular freshwater flushes to maintain the range of salinity variation and nutrient and phytoplankton supply to which oysters are adapted. Therefore, changes to rainfall due to climate change, in conjunction with sea-level rise, will deliver complex interactions with unpredictable outcomes.

2.8.1 Role of policy in response to coastal asset risks

Land management agencies will need to consider and integrate aquaculture and fisheries agency management policies where coastal infrastructure is concerned. Often aquaculture lease areas and industry priorities are identified by one governing agency, but may conflict with the neighbouring but jurisdictionally separate shoreline areas. With improved jurisdictional integration, adaptation options can be addressed – for example, inland retreat in severe cases, or protective or reinforced infrastructure as an alternative.

Technology systems that can assist in the alignment of jurisdictional boundaries are well established (e.g., geographic information systems), and this alignment of spatial information on land use, natural resources, monitoring information, regional industries and socioeconomic priorities will be central to achieving better cross-jurisdictional management. National strategies and policies to provide easily accessible and powerful spatial information tools at multiple scales are vital, with the biggest remaining barrier to achieving this being inter-organizational logistics. Coastal areas are some of the most data-rich regions on the planet, from small-scale local research to national resource management initiatives. However, little of this data is ever coordinated and extended to the relevant stakeholders. This has been recognized as a weakness and an inefficient use of what is often publically funded research and monitoring. Global and national initiatives exist for the centralization of data sources from the marine environment (e.g., Australian Ocean Data Network, http://portal.aodn.org.au), and these could be harnessed, contributed to, and applied to scenarios for adapting to climate change. The development of sea-level-rise monitoring programmes and the mapping of high-risk areas has already been undertaken in many places in the Asia Pacific (Global Education 2010; Middelmann 2007).

Another adaptation to coastal impacts on aquaculture is the translocation of industry offshore through deep-sea-based infrastructure development, such as sea cages or long lines. However, this usually implies a different form of aquaculture, requiring significant skills development. Oceans technology has developed rapidly and engineering solutions can now withstand rough offshore environments, but new policy issues will need to be tackled.

Catchment remediation initiatives are underway in some regions to improve coastal waterways, including the replanting of mangroves and riparian vegetation. In addition, commercial fishers and growers are starting to implement environmental management systems (EMS), which provide them with useful guidance for improving environmental management of their production and operation systems. Thus, as the beneficiaries of healthy coastal habitats, fisheries and aquaculture operators can become the stewards of catchments, preparing them for and protecting them from adverse climate change impacts. A consideration here is that accessible financing systems are required for many of the marine living resource industries to both improve current standards and safeguard coastal infrastructure.

2.9 Lack of Feed

A lack of feed for wild fish stocks has already been suggested as the cause of reduced productivity for a number of globally important fish species, including the Alaskan Salmon (Armstrong et al. 2005), Pacific Sardine and Northern Anchovy (Finney et al. 2002). It is thought that a lack of feed in wild stocks is an issue primarily for larval and juvenile life stages, which have evolved with the timing and place of prey production.

In relation to aquaculture of fish and other fed species, future growth is challenged by these limitations of feed, as much aquaculture feed is currently produced through the wild harvest of small, oily fish. The growth in aquaculture has been the single biggest cause of the rapid price increases of fish oil and fish meal, critical sources of lipid and protein, in the past decades (Peron et al. 2010). This is one of the reasons that small-species capture fisheries continue to grow, while other capture fisheries are stable or declining. There has been some replacement of aquaculture feed protein with cereal crops such as soybeans. However, land crop productivity is also under pressure and such protein sources are rarely nutritionally complete for marine species.

2.9.1 Role of policy in response to feed shortage

Solutions to the pressure on and supply of aquaculture feeds need to be found. Some contribution of terrestrial low-value protein, carbohydrate and lipid sources can contribute in part to aquaculture feed supplies. However, considering that land food production will likely also be under stress as a result of climate change this is not a solution. Other developments include the production of feeds from marine crop (seaweed and microalgae) production (FAO 2010a; Bolton et al. 2009), insect cultivation (Alegbeleye et al. 2012) and/or a transition in aquaculture to the cultivation of species lower on the food chain such as herbivorous fish and shellfish.

Another holistic strategy for sustainable food production from the oceans that can integrate feed production is the concept of marine agricultural systems. This system, in which China is the world leader, has come to be termed the turquoise revolution (blending blue for water-based aquaculture and green for land-based agriculture). The concept of marine agriculture is that of integrated production systems that incorporate a complementary suite of marine living resources from primary production to higher-level consumers. A key feature of this concept is that there are limits to the production of higher trophic-level species, and that aquaculture must focus on the development of lower food chain taxa.

2.10 Impacts to Human Health

The issue of human health and living marine resources is complex and inclusive of multiple disciplines. For many in the seafood sector, human health issues relate directly to acute disease outbreaks as a consequence of inadequate quality assurance programmes, coastal pollution (e.g. sewage outfalls) and toxic algal blooms (which have the potential to increase in frequency and extent with the onset of climate change). However, from a regional and governance sector perspective, chronic deficiencies in human health security caused by inadequacies in nutritional food production systems is a major concern – with most immediacy for the less socioeconomically stable nations of the Asia Pacific (Wahlqvist and Kuo 2009).

One of the most common acute human illnesses caused by seafood consumption in the tropical Asia Pacific region is ciguatera poisoning; with French Polynesia being one of the global hotspots of occurrence (Chinain et al. 2010). Ciguatera poisoning results from the consumption of tropical herbivorous reef fishes, or their predators, that have consumed toxic microalgae. Consumption of the liver and viscera of fishes increases the risk of ciguatera poisoning, which, though not always lethal (Hamilton et al. 2010), can have long-lasting impacts on human health. This has had significant economic and health consequences in developing nations, including in the Asia Pacific (Baumann et al. 2010).

The microalgae that produce the ciguatera toxin are dinoflagellates (*Gambierdiscus toxicus*). Many species of this group of phytoplankton are positively correlated with increases in sea-surface temperature (Tester et al. 2010) and ciguatera events are correlated with the Southern Oscillation Index (SOI). However, there is a further threshold for extended periods of warm sea-surface temperature in which ciguatera prevalence decreases (Llewellyn 2010). As in most cases, the effects of climate change will be variable and inconsistent across the climatic zones. Therefore, monitoring and modelling are central to making an informed response, in this case to the occurrence of algal blooms. Likely response will include education within the community (Chinain et al. 2010; Donati 2006).

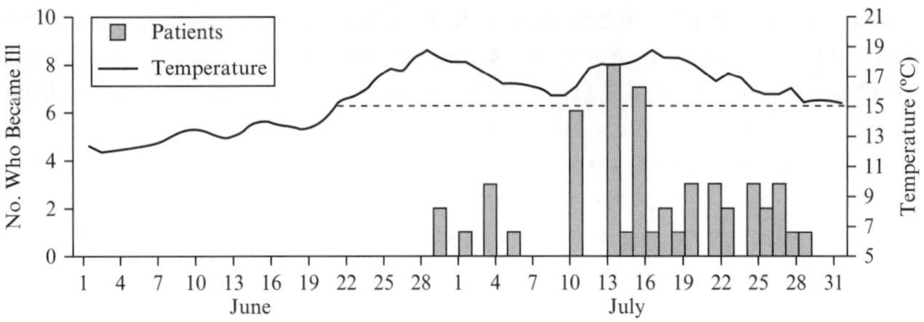

Notes: The grey dashed line denotes a theorized threshold for *Vibrio* prevalence.

Source: McLaughlin et al. (2005). Copyright © (2012) reprinted with permission from Massachusetts Medical Society.

Figure 4.5 The correlation between gastroenteritis events and rises in sea-surface temperatures using data from gastroenteritis patients identified as having eaten oysters from the same oyster farm

In temperate climates, acute diseases can also arise from HABs as well as from naturally occurring marine bacteria, with subsequent implications for shellfish handling as the climate warms. Preliminary data indicates that the prevalence of human diseases (including cholera) from seafood, particularly shellfish, may be indirectly linked to sea-surface temperature and climate change (Lobitz et al. 2000; Patz et al. 1996). McLaughlin et al. (2005) found Alaskan oyster farm areas have tracked an increase in temperature of 0.21C per annum since 1997, and that this correlates with an increase in the number of patients with gastroenteritis caused by *Vibrio paphaemolyticus*. *V. paphaemolyticus* is a naturally occurring marine bacteria that appears to become more prolific and/or more of a seafood safety issue with warmer sea temperatures (see Figure 4.5).

Of potentially greater concern than acute health impacts is global health and food security, which is dependent on ecological systems (Wahlqvist and Kuo 2009). Chronic malnutrition and diseases are a well-documented extreme outcome of poor food production systems, yet more subtle impacts on society, such as rates of coronary and metabolic disease and quality of mental health and physical development can also be attributed to the food production system. Seafood is an important component of the basic nutritional requirements of proteins and energy, but of equal importance are the trace elements and biochemical components of seafood that contribute functionally to the health, vigour and mental condition of societies (McManus et al. 2009). Therefore, global management of fisheries production needs to address the capacity of the oceans to provide for food and health security.

2.10.1 Role of policy in response to human health risks

Central to the risk of human disease linked to climate change in the oceans is monitoring both on the ground and by using remote sensing technology where suitable (Lobitz et al. 2000). Such monitoring already includes quality assurance programmes for food, such as for shellfish or fish, with both water and tissue being tested for the presence of algae, viruses or bacteria. Long-term assimilation to detect trends correlated to climate change will be important. Some new genetic technologies are also being developed to allow for more rapid and cost-effective detection of pathogens in water samples. These will assist in achieving cost-effective monitoring (Murray et al. 2011). In addition, the promotion of and investment in sustainable and nutritionally valuable seafood production technologies, particularly in aquaculture will increasingly be needed to supply adequate nutrition to future populations (World Bank 2009).

2.11 Harvest Regularity and Market Fluctuation

It is anticipated that temporary closure of fisheries or reduced fishing opportunities may result from fluctuating stock numbers (e.g., in the case of anchoveta), HABs, and increasing pollution incidents due to greater storm run-off (e.g., impacting shellfish culture), disease (e.g., impacting abalone) or higher frequencies of extreme storm events. This will reduce the number of available working days and affect the industry's operations and market opportunities. However, with some foresight, it should be possible to structure flexible schedules, although in some instances closures could be lengthy.

Extreme weather can also affect operations in the fishing and aquaculture sectors through increases in work hours during storm events and destruction of critical infrastructure to operations (e.g., ports, fishing and aquaculture gear). It is estimated that the insurance and government expenditure in response to extreme climate events in Australia has increased substantially over the past two decades (Middelmann 2007), with this figure likely set to increase in response to climate change, particularly for offshore installations and operations. In addition, offshore operations could become more hostile. Understanding offshore conditions is likely to be more important in the future, as aquaculture is moved further offshore to seek an increasingly limited wild harvest (Davies et al. 2007).

2.11.1 Role of policy in response to variable harvest and markets

A key opportunity here is to provide industries with improved environmental data for long-term and real-time trends so that industry can predict when and where closure events might occur. This is analogous to the use of meteorological data and the SOI for terrestrial farming systems, and would allow farmers to respond in real time through immediate harvest in anticipation of a closure event (e.g., rainfall and algal blooms), or by applying longer-term strategies, such as spreading the distribution of lease areas or pursuing diversification of species. Governance has a role to play here in the stimulation of monitoring programmes that deliver some predictive power, and in supporting regional and national food safety and quality assurance programmes.

Fishery and aquaculture failures are not always due to biological or productivity collapse, but may be owing to mismatches in supply and demand of markets (Hilborn et al. 2005). There are opportunities to adapt to this through seasonal balancing with competitors in the market, as occurred with the New Zealand and Western Australian rock lobster fisheries who co-function to maintain a consistent market supply despite local fluctuations (Breen and Kendrick 1997).

3. POLICY IMPLICATIONS OF IMPACTS TO MARINE LIVING RESOURCES

Presented here has been a summary of the direct and indirect effects of climate change on the marine living resource sectors of wild fisheries and aquaculture. As reviewed in this chapter, categories of impacts from climate change on marine living resources are predictable and, in many cases, already measurable. Many marine living resources are changing at the scale of species to whole habitats, with socioeconomic systems being vulnerable to these changes. Short-term industry adaptation and governance responses for policy makers to consider in light of these demonstrable and predicted changes have been discussed and are summarized in Table 4.2, many of which are common across the 11 climate change impacts identified here.

There remain challenges and opportunities for the improvement of governance systems for marine living resources and the solutions to the challenge of climate change will be as diverse as the nations that face these challenges. The urgency and types of limitations will differ dramatically across the diverse climates, industries and socioeconomic conditions of the Asia Pacific. In this chapter the policy currents and recommendations delivered are categorized and can deliver some commonality to the issues and solutions across the diverse nations and industry barriers in the Asia Pacific.

Table 4.2 Synthesis of diverse industry, governance and policy responses discussed in this chapter in relation to eleven impacts of climate change on marine living resources

Response to impacts		Production and yield	Type of yield (trophic level)	Mortality (disease & stress)	Pests & fouling	Seed supply/recruitment	Species shifts	Genetic diversity	Coastal assets	Lack of feed	Human health	Harvest & market fluctuation
	Direct industry impacts											
Industry management	Spatial management											
	Species management											
	Bio-security protocols											
	Selective breeding											
	Hatchery development											
	Transitional technology											
	Diversification											
Monitoring & research	Industry											
	Climate and natural resources											
	Integration & communication											
Vertical integration	Market structures											
	Industry management											
	Diversification of markets											
Policy	Management facilitation											
	Carrying capacity											
	Resource allocation & fisheries											
	Communication											
	Responding to monitoring											
	Industry incentives/compliance											
	Industry transition											
	Financing systems											

REFERENCES

ABARE (2007), *Fishery Economic Status Report 2007*, Canberra: Australian Bureau of Agricultural and Resource Economics , Ch. 3.

ABARE (2009), *Australian Fisheries Statistics 2008*, Canberra: Australian Bureau of Agricultural and Resource Economics.

Alegbeleye, W.O., S.O. Obasa, O.O. Olude, K. Otubu and W. Jimoh. (2012), 'Preliminary evaluation of the nutritive value of the variegated grasshopper (*Zonocerus variegatus* L.) for African catfish *Clarias gariepinus* (Burchell. 1822) fingerlings', *Aquaculture Research*, **43** (3), 412–420.

Alvial, A.A. (2010), 'Health and environmental regulations in Chile move forward', *The Chilean Aquaculture Observer*, **1**.

Anestis, A., H.O. Portner, D. Karagiannis, P. Angelidis, A. Staikou and B. Michaelidis (2010), 'Response of *Mytilus galloprovincialis* (L.) to increasing seawater temperature and to marteliosis: metabolic and physiological parameters', *Comparative Biochemistry and Physiology*, Part A **156** (1), 57–66.

APFIC Asia Pacific Fishery Commission (2010), *Balancing the Needs of People and Ecosystems in Fisheries and Aquaculture Management in the Asia Pacific*, 3rd Regional Consultative Forum Meeting, Jeju Island, Republic of Korea.

Aquenal Pty Ltd (2008), *National Control Plan for the Japanese Seaweed or Wakame* (Undaria pinnatifida), The National System for the Prevention and Management of Marine Pest Incursions, Canberra.

Armstrong, J.L., J.L. Boldt, A.D. Cross, J.H. Moss and N.D. Davis (2005), 'Distribution, size, and interannual, seasonal and diel food habits of northern Gulf of Alaska juvenile pink salmon, *Oncorhynchus gorbuscha*', *Deep Sea Research*, **II** (52), 247–265.

Arnold, K.E., H.S. Findlay, J.I. Spicer, C.L. Daniels and D. Boothroyd (2009), 'Effect of CO2-related acidification on aspects of the larval development of the European lobster, *Homarus gammarus* (L.)', *Biogeosciences*, **6**, 1747–1754.

Atrill, M.J., J. Wright and M. Edwards (2007), 'Climate-related increases in jellyfish frequency suggest a more gelatinous future for the North Sea', *Limnology and Oceanography*, **52**, 480 485.

Banks, S.C., S.D. Ling, C.R. Johnson, M.P. Piggott, J.E. Williamson and L.B. Beheregaray (2010), 'Genetic structure of a recent climate change-driven range extension', *Molecular Ecology*, **19** (10), 2011–2024.

Battaglene, S., C. Carter and A.J. Hobday (2008), *Scoping Study into the Adaptation of the Tasmanian Salmonid Aquaculture Industry to Potential Impacts of Climate Change*, Tasmanian Aquaculture and Fisheries Institute.

Baumann, F., M.-B. Bourrat and S. Pauillac (2010), 'Prevalence, symptoms and chronicity of ciguatera in New Caledonia: results from an adult population survey conducted in Noumea during 2005', *Toxicon*, **56** (5), 662–667.

Bax, N., A. Williamson, M. Aguero, E.G. Gonzalez and W. Geeves (2003), 'Marine invasive alien species: a threat to global biodiversity', *Marine Policy*, **27**, 313–323.

Beardal, J. and J.A. Raven (2004), 'Potential effects of global change on microalgal photosynthesis, growth and ecology', *Phycologia*, **43**, 26–40.

Beaugrand, G., M. Edwards and L. Legendre (2010), 'Marine biodiversity, ecosystem functioning, and carbon cycles', *Proceedings of the National Academy of Sciences (USA)*, **107** (22), 10120–10124.

Behrenfeld, M.J., R.T. O'Malley, D.A. Siegel, C.R. McClain, J.L. Sarmiento, G.C. Feldman, A.J. Milligan, P.G. Falkowski, R.M. Letelier and E.S. Boss (2006), 'Climate-driven trends in contemporary ocean productivity', *Nature*, **444** (7120), 752.
Bell, J.D., T.J.H. Adams, J.E. Johnson, A.J. Hobday and A. Sen Gupta (2011), 'The Pacific Community, fisheries, aquaculture and climate change: an introduction', in J.D. Bell (ed.), *Vulnerability of Fisheries and Aquaculture in the Pacific to Climate Change*, New Caledonia: Secretariat of the Pacific Community, Noumea.
Bell, J.D., D.M. Bartley, K. Lorenzen and N.R. Loneragan (2006), 'Restocking and stock enhancement of coastal fisheries: potential, problems and progress', *Fisheries Research*, **80**, 1 8.
Bolton, J.J., D.V. Robertson-Anderson, D. Shuuluka and L. Kandjengo (2009), 'Growing *Ulva* (Chlorophyta) in integrated systems as a commercial crop for abalone feed in South Africa: a SWOT analysis', *Journal of Applied Phycology*, **21**, 575–583.
Breen, P.A. and T.H. Kendrick (1997), 'A fisheries management success story: the Gisborne, New Zealand, fishery for red rock lobsters (*Jasus edwardsii*)', *Marine and Freshwater Research*, **1997** (48), 1103–1110.
Byrne, M., M. Ho, E. Wong, N. Soars, P. Selvakumaraswamy, S. Dworjanyn and A.R. Davis (2010), *Unshelled Abalone and Corrupted Urchins, Development in a Changing Ocean*, Australian Marine Science Association: New Waves in Marine Science, Wollongong, Australia: AMSA.
Camara, M.D., S. Evans and C.J. Langdon (2008), 'Parental relatedness and survival of pacific oysters from a naturalized population', *Journal of Shellfish Research*, **27** (2), 323–336.
Canu, D.M., C. Solidoro, G. Cossarini and F. Giorgi (2010), 'Effect of global change on bivalve rearing activity and the need for adaptive management', *Climate Research*, **42**, 13–26.
Chassot, E., S. Bonhommeau, N.K. Dulvy, F. Mélin, R. Watson, D. Gascuel and O. Le Pape (2010), 'Global marine primary production constrains fisheries catches', *Ecology Letters*, **13** (4), 495–505.
Cheung, W.W.L., V.W.Y. Lam, J.L. Sarmiento, K. Kearney, R. Watson, D. Zeller and D. Pauly (2010), 'Large-scale redistribution of maximum fisheries catch potential in the global ocean under climate change', *Global Change Biology*, **16**, 24–35.
Chinain, M., H.T. Darius, A. Ung, M.T. Fouc, T. Revel, P. Chruchet, S. Pauillac and D. Laurent (2010), 'Ciguatera risk management in French Polynesia: the case study of Raivavae Island (Australes Archipelago)', *Toxicon*, **56** (5), 674–690.
Church, J.A., J.R. Hunter, K.L. McInnes and N.J. White (2006a), 'Sea-level rise around the Australian coastline and the changing frequency of extreme sealevel events', *Australian Meterological Magazine*, **55**, 253–260.
Church, J.A., N.J. White and J.R. Hunter (2006b), 'Sea-level rise at tropical Pacific and Indian Ocean islands', *Global and Planetary Change*, **53**, 155–168.
Corrales, R.A. and J.L. Maclean (1995), 'Impacts of harmful algae on seafarming in the Asia-Pacific areas', *Journal of Applied Phycology*, **7** (2), 151–162.
Coutts, A.D.M. and T.J. Dodgshun (2007), 'The nature and extent of organisms in vessel sea-chests: a protected mechanism for marine bioinvasions', *Marine Pollution Bulletin*, **54**, 875–886.

Crockford, M. (2007), *Partial Characterisation of Pilchard Herpesvirus and the Associated Disease in Pilchards*, Melbourne: Murdoch University.

Cullis-Suzuki, S. and D. Pauli (2010), 'Failing the high seas: a global evaluation of regional fisheries management organisations', *Marine Policy*, **34** (5), 1036–1042.

Davies, A.J., J.M. Roberts and J. Hall-Spencer. (2007), 'Preserving deep-sea natural heritage: emerging issues in offshore conservation and management', *Biological Conservation* **138** (3–4): 299–312

De Oliveira, L.R., D. Meyer, J. Hoffman, P. Majluf and J.S. Morgante (2009), 'Evidence of a genetic bottleneck in an El Niño affected population of South American fur seals, *Arctocephalus australis*', *Journal of the Marine Biological Association of the United Kingdom*, **89** (8), 1717–1725.

DeMartini, E.E., F.A. Parrish and J.D. Parrish (1996), 'Interdecadal change in reef fish populations at French Frigate Shoals and Midway Atoll, Northwestern Hawaiian Islands: statistical power and retrospect', *Bulletin of Marine Science*, **58** (3), 804–825.

Donati A.C. (2006), *Ciguatera Poisoning*, Fishnote No 41, Darwin, Australia.

Done, T.J. and R.E. Reichelt (1998), 'Integrated coastal zone and fisheries exosystem management: generic goals and performance indices', *Ecological Applications*, **8** (1 suppl.), S110–S118.

Donelson, J.M., P.L. Munday, P.L. McCormick, N.W. Pankhurst and P.M. Pankhurst (2010), 'Effects of elevated water temperature and food availability on the reproductive performance of a coral reef fish', *Marine Ecology Progress Series*, **401**, 233–243.

Doney, S.C., V. Fabry, R. Feely and J.A. Kleypas (2009), 'Ocean acidification: the other CO2 problem', *Annual Review of Marine Science*, **1**, 169–192.

Dudgeon, C.L., D. Broderick and J.R. Ovenden (2009), 'IUCN classification zones concord with, but underestimate, the population genetic structure of the zebra shark Stegostoma fasciatum in the Indo-West Pacific', *Molecular Ecology*, **18** (2), 248–261.

Fabry, V.J., B.A. Seibel, R.A. Feely and J.C. Orr (2008), 'Impacts of ocean acidification on marine fauna and ecosystem processes', *ICES Journal of Marine Science: Journal du Conseil*, **65** (3), 414–432.

FAO (2008a), *The State of the World Fisheries and Aquaculture – 2008*, Rome: United Nations Fisheries and Agricultural Organization.

FAO (2008b), *Yearbook of Fishery Statistics Summary Tables*, Rome: United Nations Fisheries and Agricultural Organistaion.

FAO (2009), *Climate Change Implications for Fisheries and Aquaculture*, FAO Fisheries and Aquaculture Technical Paper, 530. K. Cochrane, C. De Young, D. Soto and T. Bahri (eds), Rome: United Nations Fisheries and Agricultural Organization.

FAO (2010a), *Algae-based Biofuels: Applications and Co-products*, Environment and Natural Resources Management Working Paper, Rome: United Nations Food and Agricultural Organization

FAO (2010b), *The State of the World Fisheries and Aquaculture – 2010*, Rome: United Nations Fisheries and Agricultural Organization.

Feely, R., C. Sabine, K. Lee, W. Berelson, J. Keypas, V. Fabry and F. Millero (2004), 'Impact of anthropogenic CO2 on the CaCO3 system in the oceans', *Science*, **305**, 362–366.

Finney, B., I. Gregory-Eaves, M. Douglas and J. Smol (2002), 'Fisheries

productivity in the northeastern Pacific Ocean over the past 2,200 years', *Nature*, **416**, 729–733.
Fossa, J.H., R. Bellerby and T. Jakobsen (2008), *Consequences of Ocean Acidification for Fisheries*, The Ocean in a High CO2 World, Monaco, 6–9 October.
Global Education (2010), *South Pacific Sea Level and Climate Monitoring Project*, accessed 13th August 2012 at: http://www.bom.gov.au/pacificsealevel/index.shtml.
Goldstien, S.J., D.R. Schiel and N.J. Gemmell (2010), 'Regional connectivity and coastal expansion: differentiating pre-border and post-border vectors for the invasive tunicate Styela clava', *Molecular Ecology*, **19** (5), 874–885.
Hallegraeff, G.M. (2010), 'Ocean climate change, phytoplankton community responses, and harmful algal blooms: a formidable predictive challenge', *Journal of Phycology*, **46** (2), 220–235.
Hamilton, B., N.J. White, G. Shaw, G. Eaglesham, M.R. Moore and R.J. Lewis (2010), 'Human fatality associated with Pacific ciguatoxin contaminated fish', *Toxicon*, **56** (5), 668–673.
Haupt, T.M., C.L. Griffiths, T.B. Robinson and A.F.G. Tonin (2010), 'Oysters as vectors of marine aliens, with notes on four introduced species associated with oyster farming in South Africa', *African Zoology*, **45** (1), 52–62.
Hiddink, J.G. and R. Ter Hofstede (2008), 'Climate induced increases in species richness of marine fishes', *Global Change Biology*, **14** (3), 453–460.
Hilborn, R., J.M.L. Orensanz and A.M. Parma (2005), 'Institutions, incentives and the future of fisheries', *Philosophical Transcripts of the Royal Society B: Biological Sciences*, **360** (1453), 47–57.
Hobday, A.J. and E.S. Poloczanska (2010), 'Marine fisheries and aquaculture', in C. Stokes and M. Howden (eds), *Adapting Agriculture to Climate Change*, Collingwood: CSIRO Publishing.
Johannes, R.E. (2002), 'The renaissance of community-based marine resource management in oceania', *Annual Review of Ecological Systems*, **33**, 317–340.
Johnson, J.E. and D.J. Welch (2010), 'Marine fisheries management in a changing climate: a review of vulnerability and future options', *Reviews in Fisheries Science*, **18** (1), 106–124.
Klyashtorin, L.B. (2001), *Climate Change and Long-Term Fluctuations of Commercial Catches – The Possibility of Forecasting*, FAO Fisheries Technical Paper 410, Rome: United Nations Fisheries and Agricultural Organization.
Kompas, T. (2008), 'Managing the Gulf of Papua prawn fishery: sustainability, maximum returns and cooperation between commercial fishing and indigenous fishing communities', *Pacific Economic Bulletin*, **23** (1), 29–38.
Krkošek, M. (2010), 'Review: host density thresholds and disease control for fisheries and aquaculture', *Aquaculture Environment Interactions*, **1**, 21–32.
Kronen, M., A. Vunisea, F. Magron and B. McArdle (2010), 'Socio-economic drivers and indicators for artisanal coastal fisheries in Pacific Island countries and territories and their use for fisheries management strategies', *Marine Policy*, **34** (6), 1135–1143.
Lang, R.P., C.J. Bayne, M.D. Camara, C. Cunningham, M.J. Jenny and C.J. Langdon (2009), 'Transcriptome profiling of selectively bred pacific oyster Crassostrea gigas families that differ in tolerance of heat shock', *Marine Biotechnology*, **11** (5), 650–668.

Larsen, T., S.H. Thilsted, K. Kongsbak and M. Hansen (2000), 'Whole small fish as a calcium source', *British Journal of Nutrition*, **83**, 191–196.
Last, P.R., W.T. White, D.C. Gledhill, A.J. Hobday, A.J. Brown, G.J. Edgar and G. Pecl (2011), 'Long-term shifts in abundance and distribution of a temperate fish fauna: a response to climate change and fishing practices', *Global Ecology and Biogeography*, **20** (1), 58–72.
Lessios, H.A. and D.R. Robertson (2006), 'Crossing the impassable: Genetic connections in 20 reef fishes across the Eastern Pacific barrier', *Proceedings of the Royal Society B: Biological Sciences*, **273** (1598), 2201–2208.
Levitan, O., G. Rosenberg, I. Setlik, E. Setlikova, E. Grigel, J. Klepetar, O. Prasil and I. Berman-Frank (2007), 'Elevated CO_2 enhances nitrogen fixation and growth in the marine cyanobacterium *Trichodesmium*', *Global Change Biology*, **13**, 531–538.
Lind, C.E., B.S. Evans, J.J.U. Taylor and D.R. Jerry (2007), 'Population genetics of a marine bivalve, Pinctada maxima, throughout the Indo-Australian Archipelago shows differentiation and decreased diversity at range limits', *Molecular Ecology*, **16** (24), 5193–5203.
Ling, S.D., C.R. Johnson, S. Frusher and C.K. King (2008), 'Reproductive potential of a marine ecosystem engineer at the edge of a newly expanded range', *Global Change Biology*, **14** (4), 907–915.
Llewellyn, L.E. (2010), 'Revisiting the association between sea surface temperature and the epidemiology of fish poisoning in the South Pacific: Reassessing the link between ciguatera and climate change', *Toxicon*, **56** (5), 691–697.
Lobitz, B., L. Beck, A. Huq, B. Wood, G. Fuchs, A.S.G. Faruque and R. Colwell (2000), 'Climate and infectious disease: use of remote sensing for detection of *Vibrio cholerae* by indirect measurement', *PNAS*, **97** (4), 1438–1443.
Lopez, M.D., M.U. Alcocer and P.D. Jaimes (2010), 'Phylogeography and historical demography of the Pacific Sierra mackerel (*Scomberomorus sierra*) in the Eastern Pacific', *BMC Genetics*, **11**, 34.
Lough, J.M. (2009), 'Temperature', in *A Marine Climate Change Impacts and Adaptation Report Card for Australia 2009*, E.S. Poloczanska, A.J. Hobday and A.J. Richardson (eds), NCCARF.
Lovatelli, A. (2004), *Advances in Sea Cucumber Aquaculture and Management*, FAO Fisheries Technical Paper No. 463, C. Conand, S. Purcell, S. Uthicke, J.-F. Hamel and A. Mercier (eds), Rome: Food and Agricultural Organization of the United Nations.
Lymer, D., S. Funge-Smith, J. Clausen and W. Miao (2008), *Status and Potential of Fisheries and Aquaculture in Asia and the Pacific 2008*. APFIC Asia Pacific Fisheries Commission, FAO Regional Office for Asia and the Pacific.
McLaughlin, J.B., A. DePaola, C.A. Bopp, K.A. Martinek, N.P. Napolilli, C.G. Allison, S.L. Murray, B.S. Thompson, M.M. Bird and J.P. Middaugh (2005), 'Outbreak of *Vibrio parahaemolyticus* gastroenteritis associated with Alaskan oysters', *The New England Journal of Medicine*, **353** (14), 1463–1470.
McManus, A., J. Howieson and J. Nicholson (2009), *Review of Literature and Resources Relating to the Health Benefit of Regular Consumption of Seafood as Part of a Healthy Diet*, Perth: Centre of Excellence Science, Seafood and Health, Curtin Health Innovation Research Institute, Curtin University of Technology.
Middelmann, M.H. (2007), *Natural Hazards in Australia: Identifying Risk Analysis Requirements*, Canberra: Geoscience Australia.

Mollmann, C., B. Muller-Karulis, G. Kornilovs and M.A. St John (2008), 'Effects of climate and overfishing on zooplankton dynamics and ecosystem structure: regime shifts, trophic cascade, and feedback loops in a simple ecosystem', *ICES Journal of Marine Science: Journal du Conseil*, **65** (3), 302–310.

Murray, S.A., M. Wiese, A. Stuken, S. Brett, R. Kellmann, G. Hallegraeff and B.A. Neilan (2011), 'A quantitative molecular assay based on the gene sxtA to identify saxitoxin-producing harmful algal blooms in marine waters', *Applied Environmental Microbiology*, **77** (19), 7050–7057.

NCCARF (2009), *Report Card of Marine Climate Change for Australia*. E.S. Poloczanska, A.J. Hobday and A.J. Richardson (eds), NCCARF.

Niklitschek, E.J., D.H. Secor, P. Toledo, A. Lafon and M. George-Nascimento (2010), 'Segregation of SE Pacific and SW Atlantic southern blue whiting stocks: integrating evidence from complementary otolith microchemistry and parasite assemblage approaches', *Environmental Biology of Fishes*, 1–15.

Orr, J., V. Fabry, O. Aumont, L. Bopp, S.C. Doney, R.A. Feely, A. Hnanadeikan, N. Gruber, A. Ishida, F. Joos, R.M. Key, K. Lindsay, E. Maier-Reimer, R. Matear, P. Monfray, A. Mouchet, R.G. Naijar, G.K. Plattner, K.B. Rodgers, C.L. Sabine, J.L. Sarmiento, R. Schlitzer, R.D. Slater, I.J. Totterdell, M.F. Weirig, Y. Yamanaka and A. Yool (2005), 'Anthropogenic ocean acidification over the twenty-first century and its impact on calcifying organisms', *Nature*, **437** (7059), 681–686.

Ospina-Ålvarez, N. and F. Piferrer (2008), 'Temperature-dependent sex determination in fish revisited: prevalence, a single sex ratio response pattern, and possible effects of climate change', *PLoS ONE*, **3** (7).

Ovenden, J.R., T. Kashiwagi, D. Broderick, J. Giles and J. Salini (2009), 'The extent of population genetic subdivision differs among four co-distributed shark species in the Indo-Australian archipelago', *BMC Evolutionary Biology*, **9** (40).

PaCFA (2009), *Fisheries and Aquaculture in our Changing Climate*, Global Partnership Climate Fisheries and Aquaculture, Policy Brief.

Palero, F., P. Abello, E. Macpherson, C.A. Matthee and M. Pascual (2010), 'Genetic diversity levels in fishery-exploited spiny lobsters of the genus *Palinurus* (Decapoda: Achelata)', *Journal of Crustacean Biology*, **30** (4), 658–663.

Parker, L., P. Ross, W. O'Connor and D. Raftos (2010), *The Effects of Ocean Acidification and Temperature on Oysters and the Potential of Selective Breeding To Ameliorate Climate Change*, AMSA 2010 – New Waves in Marine Science, Wollongong: Australian Marine Science Association.

Patz, J.A., P.R. Epstein, T.A. Burke and J.M. Balbus (1996), 'Global climate change and emerging infectious diseases', *Journal of the American Medical Association*, **275** (3), 217–223.

Pauly, D., V. Christensen, J. Dalsgaard, R. Froese and F. Torres Jr. (1998), 'Fishing down marine food webs', *Science*, **279**, 860–863.

Pecl, G., S. Frusher, C. Gardner, M. Haward, A.J. Hobday, S. Jennings, M. Nirsey-Bray, A.R. Punt, H. Revill and I. van Putten (2009), *The East Coast Tasmanian Rock Lobster Fishery: Vulnerability to Climate Change and Adaptation Response Options*', Report to the Department of Climate Change, Australia.

Peron, G., J.-F. Mittaine and B. Le Gallic (2010), 'Where do fish meal and fish oil products come from? An analysis of the conversion ratios in the global fishmeal industry', *Marine Policy*, **34** (4), 815–820.

Portner, H.O. and R. Knust (2007), 'Climate change affects marine fishes through oxygen limitation of thermal tolerance', *Science*, **315**, 95–97.

Richardson, A.J., A. Bakun, G.C. Hays and M.J. Gibbons (2009), 'The jellyfish joyride: causes, consequences and management responses to a more gelatinous future', *Trends in Ecology and Evolution*, **24** (6), 312–322.
Riebesell, U., I. Zondervani, B. Rost, P.D. Tortell, R. Zeebe and F.M.M. Morel (2000), 'Reduced calcification of marine plankton in response to increased atmospheric CO2', *Nature*, **407**, 364–367.
Rigos, G., K. Bitchava and I. Nengas (2010), 'Antibacterial drugs in products originating from aquaculture: assessing the risks to public welfare', *Mediterranean Marine Science*, **11** (1), 33–41.
Shackell, N.L., K.T. Frank, J.A.D. Fisher, B. Petrie and W.C. Leggett (2010), 'Decline in top predator body size and changing climate alter trophic structure in an oceanic system', *Proceedings of the Royal Society of London*, **277** (1686), 1353–1360.
Shepard, S., B. Beukers-Stewart, J.G. Hiddink, A.R. Brand and M.J. Kaiser (2010), 'Strengthening recruitment of exploited scallops *Pecten maximus* with ocean warming', *Marine Biology*, **157** (1), 91–97.
Simms, A.D. and C.D. Woodroffe (2008), *Scoping Study of Climate Change in the Southern Rivers Region: Focus on Natural, Built, Coastal and Marine Systems*, Wollongong: Southern Rivers Catchment Management Authority, GeoQuEST Research Centre, School of Earth and Environmental Sciences, University of Wollongong.
Somero, G.N. (2010), 'The physiology of climate change: how potentials for acclimatization and genetic adaptation will determine "winners" and "losers"', *Journal of Experimental Biology*, **213** (6), 912–920.
Subasinghe, R.P., S.E. McGladdery and B.J. Hill (2004), *Surveillance and Zoning for Aquatic Animal Diseases*, FAO Fisheries Technical Paper No 451, Rome: United Nations Fisheries and Agricultural Organization.
Sutherland, K.P., J.W. Porter and C. Torres (2004), 'Disease and immunity in Caribbean and Indo-Pacific zooxanthellate corals', *Marine Ecology Progress Series*, **266**, 273–302.
Tester, M. and P. Langridge (2010), 'Breeding technologies to increase crop production in a changing world', *Science*, **327** (5967), 818–822.
Tester, P.A., R.L. Feldman, A.W. Nau, S.R. Kibler and R.W. Litaker (2010), 'Ciguatera fish poisoning and sea surface temperatures in the Caribbean Sea and the West Indies', *Toxicon*, **56** (5), 698–710.
Tomaru, Y., Z. Kawabata and S. Nakano (2001), 'Mass mortality of the Japanese pearl oyster *Pinctada fucata martensii* in relation to water temperature, chlorophyll a and phytoplankton composition', *Diseases of Aquatic Organsisms*, **44**, 61–68.
Uki, N. (2006), 'Stock enhancement of the Japanese scallop *Patinopecten yessoensis* in Hokkaido', *Fisheries Research*, **80**, 62–66.
Uye, S.-I. (2008), 'Blooms of the giant jellyfish *Nemopilema nomurai*: a threat to the fisheries sustainability of the East Asian Marginal Seas', *Plankton and Benthos Research*, **3** (suppl.), 125–131.
WA Fisheries (2010), *Western Rock Lobster Settlement Trends*, Western Australia Department of Fisheries, accessed 26th July 2012 at: http://www.fish.wa.gov.au.
Wahlqvist, M.L. and K. Kuo (2009), 'Securing health through food systems: an initiative of the nutrition consortium of the National Health Research Institutes in Taiwan and Asia Pacific regional partners as a network', *Asia Pacific Journal of Clinical Nutrition*, **18** (4), 472–479.

Watson, S.A., P.C. Southgate, P.A. Tyler and L.S. Peck (2009), 'Early larval development of the Sydney Rock Oyster *Saccostrea glomerata* under near-future predictions of CO_2-driven ocean acidification', *Journal of Shellfish Research*, **28** (3), 431–437.

Wianans, A.K. and J.E. Purcell (2010), 'Effects of pH on asexual reproduction and statolith formation of the scyphozoan, *Aurelia labiata*', *Hydrobiologia*, **645** (1), 39–52.

Wongsanga, P. and S. Vichitlekarn (2010), 'Advancing the promotion of One Village One Fisheries Product (FOVOP) in the ASEAN region: issues and challenges for future cooperation', *Fish for the People: a Special Publication for the Promotion of Sustainable Fisheries for Food Security in the ASEAN Region*, **8** (2).

World Bank (2009), 'Managing land and water to feed nine billion people and protect natural ecosystems', *World Development Report 2010: Development and Climate Change*, Washington DC.

Yan, X.H., S.Y. Ma and Y. Aruga (2007), *Selection of a High-Temperature Resistant Strain of Porphyra Haitanensis (Rhodophyta)*, XIXth International Seaweed Symposium: Science and Technology for Traditional and Modern Utilization, Kobe, Japan.

Yang, L.H. and V.H.W. Rudolf (2010), 'Phenology, ontogeny and the effects of climate change on the timing of species interactions', *Ecology Letters*, **13**, 1–10.

5. Climate change: Antarctica and the Southern Ocean, science, law and policy

Marcus Haward*

Human-induced emissions of greenhouse gases are considered to contribute to climate change through global warming (IPCC 2009). This warming contributes to rises in ocean temperature, and is considered a primary factor in increased climate variability, including the likelihood of more frequent 'extreme events' such as cyclones and storms that directly impact on coastal areas. Concomitant changes in the ocean's chemistry have longer-term and likely irreversible impacts. Antarctica and the Southern Ocean, including the atmosphere above, are key components of the global climate system. The Southern Ocean connects the world's great ocean basins, and is a source of physical and chemical processes arising from the interaction of wind, ice and ocean.

This chapter focuses first on the critical role of the Southern Ocean and Antarctica in global and regional climates through the interaction of elements of the cryosphere (those parts of the world that are covered by ice and snow). These elements of the cryosphere provide the impetus for the thermohaline circulation (the movement of water due to differences in temperature and salinity), which plays a major role in global ocean circulation; although it is important to note that similar processes occur in the Arctic and the North Atlantic. The second part of the chapter examines the regimes and instruments that govern the Southern Ocean, focusing on the impact of climate change on the operation on these regimes. The role of science and the question of regime intersection are key areas of concern and are thus addressed in a short examination of issues surrounding ocean fertilization. The paper concludes with some observations on the influence of the intersection between science, law and policy in addressing climate change.

The thermohaline circulation is driven by differences in the density of water:

> Sea water density depends on temperature and salinity . . . salinity and temperature differences arise from heating/cooling at the sea surface and from the surface freshwater fluxes (evaporation and sea ice formation enhance salinity; precipitation, runoff and ice-melt decrease salinity). (Rahmstorf 2006, p. 1)

This leads to considerable movement of seawater. This circulation, in addition to wind-driven currents and tides, provides the major mechanisms by which heat is transferred from the equator to the poles. The generation of cold, saline Antarctic 'bottom water' from the seasonal pattern of ice freeze and thaw provides the 'engine room' for what is termed the 'overturning circulation'. These processes have been described as the 'flywheel' driving a significant part of the world's ocean circulation (Steffen 2006).

The winter–summer seasonal pattern of sea-ice formation and loss, more than doubling the ice-covered area of Antarctica, is therefore a driver in the climate system, with important links to the region's ecosystem (ACE CRC 2009b). The Southern Ocean is also a major store or sink of the world's carbon, and an integral part of the carbon cycle, where atmospheric carbon dioxide (CO_2) is absorbed by chemical and biological processes. Ecosystem change, too, will have impacts and may mean that resources of the Southern Ocean and Antarctica become more attractive due to scarcity or competition in other areas of the world. In addition, 'in a worst-case scenario of enhanced global warming, the major impacts will most likely be felt in the polar regions first' (Jabour and Haward 2009, p. 102).

Climate change also impacts the regimes and instruments that provide a framework for management of Antarctica and the Southern Ocean. These regimes and instruments include, *inter alia*, the Antarctic Treaty, the Protocol to the Antarctic Treaty on Environmental Protection (Madrid Protocol), and the Convention on the Conservation of Antarctic Marine Living Resources (CCAMLR), which provides a framework for the management of the distinctive environment, resources and ecosystem south of the Antarctic polar front – collectively known as the Antarctic Treaty System (ATS). In addition to the major 'framework' conventions – the United Nations Convention on the Law of the Sea (LOSC) and the United Nations Framework Convention on Climate Change (UNFCCC) – additional instruments including the Convention on Biological Diversity (CBD), the Convention on Migratory Species of Wild Animals (CMS), the Convention on International Trade of Endangered Species of Wild Fauna and Flora (CITES), the International Convention for the Regulation of Whaling (ICRW), the International Convention for the Prevention of Pollution from Ships (MARPOL 73/78) and the Convention/Protocol on the Prevention of Marine Pollution by Dumping of Wastes and Other

Matter (London Convention and London Protocol) are also influential. The intersection, too, of these framework conventions and instruments with the ATS provides important issue agendas for future development of Antarctic and Southern Ocean regimes.

Climate change also acts to sharpen the focus on the effectiveness of existing arrangements and opens up new and emergent issues, enhancing the need for greater understanding of the scope and magnitude of change. While science has long been a driver for nations' engagement in and involvement with Antarctica (for example, the 'Heroic Era' expeditions led by Scott and Mawson had scientific as well as exploration imperatives), this relationship has developed and broadened over the past 50 years. International collaboration in Antarctica during the International Geophysical Year (IGY) in 1957–58, at the height of the cold war, provided powerful support for the negotiation of the Antarctic Treaty in 1959 (Jabour and Haward 2009; see also Haward et al. 2006). Antarctic science falls into two, related forms of activity: science that can only be done in the region – for example, ice-core drilling to gain a record of past climate; or science that is best done in Antarctica due to its unique environment – for example, upper atmospheric physics and astronomy (AAD 2002). Contemporary Antarctic research is increasingly important in earth systems science, addressing 'big questions' related to global climate and global change.

Antarctic science programmes continued to expand through the 1980s and 1990s to focus on emerging earth systems science questions, driven by increased interest in global climate change. Major programmes in Antarctic marine living resources, atmospheric sciences, geosciences, glaciology and oceanography have been developed, extending earlier science programmes in cosmic ray physics and astronomy, which had developed from the first work associated with the IGY (AAD 2002). In Australia, for example, the Australian government's Antarctic Science Advisory Committee (ASAC) commissioned an external evaluation of the Antarctic science programme in 2002–2003. This led to the release of a five-year science programme that refocused scientific research into four multi-disciplinary programmes: ice, oceans, atmosphere and climate; Southern Ocean ecosystems; adaptation to environmental change; and the impact of human activity in Antarctica (Haward et al. 2006). A new science programme began in 2011–12, encompassing three themes: climate processes and change; terrestrial and near-shore ecosystems; and Southern Ocean ecosystems. This will provide continuity in research efforts. A fourth theme, frontier science, provides opportunities for research outside these themes (Stoddart and Haward 2011, p. 151). This work has also centred on research to understand the role of the

Southern Ocean in global climate, with increasing use of new technologies and equipment (Bork et al. 2008); for example, moorings, floats and gliders, providing data and knowledge, and improving understanding of the rate and extent of physical and biological changes to the oceans. Researchers are also attaching small sensors to elephant seals, which can dive to 1500 m in depth. These sensors collect data on temperature, depth and salinity and transmit these data to satellites; they are proving to be exceptionally useful sources of data. The development of the Argo programme of autonomous floats deployed in the world's oceans and transmitting to the Jason Satellite (Argo 2010; Rintoul 2010) has dramatically increased the rate of capture of oceanographic data. The deployment of Argo floats in the Southern Ocean has provided more data in the period 2003–2010 than in the preceding 50 years of shipborne scientific observations.

1. CLIMATE CHANGE, ANTARCTICA AND THE SOUTHERN OCEAN

The Antarctic continent contains large volumes of ice within its ice sheets and ice shelves, with ice over four kilometres thick in places. The volume of ice in Antarctica is significant, with the region estimated to be storing approximately 70 per cent of the world's freshwater (ACE CRC 2009a). If this ice melted, the global sea level would be around 60 metres above present levels (ACE CRC 2009a). Given the volume of ice involved, and the temperature changes required to melt it, this scenario is unlikely, yet plausible on millennial time scales with no mitigation of global warming. Antarctic ice-sheet-melt loss may contribute to sea-level rise over the coming century. While parts of Antarctica have been subject to increased melt, others have not. Near-surface temperatures in Antarctica have shown continent-wide warming trends since 1957 (Steig et al. 2009), with significant warming over the Antarctic Peninsula and much of West Antarctica. The overall continent-wide trends are partly offset by autumn cooling in East Antarctica. Major continent-wide warming is also seen higher in the atmosphere (in the mid-troposphere) in winter (Turner et al. 2006).

As noted above, the Southern Ocean is the engine room for global climate through complex processes by which sea-ice formation releases cold saline water that sinks, forming what is termed Antarctic Bottom Water. This bottom water is a critical component in driving the pattern of global current flow known as the overturning circulation. In addition, the circumpolar movement of water around Antarctica provides distinct physical and biological oceanographic conditions, with this current forming the Antarctic polar front – previously termed the Antarctic convergence.

The oceans also provide a major sink for atmospheric carbon, by absorption of CO_2 through physical and biological processes (the physical and biological pumps) that are part of the carbon cycle (ACE CRC 2008a). In the physical pump process, atmospheric CO_2 is dissolved into seawater and transported to the deep ocean through broader global circulation patterns (ACE CRC 2008b). These physical processes are significant, with the oceans absorbing and removing ~2 gigatonnes (GT) of carbon annually from a total of ~7 GT of carbon emitted by anthropogenic processes each year (ACE CRC 2008b). In the biological pump process, CO_2 is absorbed by small microscopic marine plants (phytoplankton) through photosynthesis. The phytoplankton is then consumed by marine animals. Other marine plants and animals also capture CO_2 and some of this CO_2 sinks to the seafloor when these organisms die (ACE CRC 2008b). The action of the biological pump, and the possibility of increasing phytoplankton blooms in areas of the oceans (such as the Southern Ocean) that are low in nutrients such as iron, have encouraged experiments in ocean fertilization (ACE CRC 2008b, pp. 4–5). While fertilization experiments (using iron and other nutrients, such as urea) have successfully induced phytoplankton blooms – with the bloom established by the Southern Ocean Iron Release Experiment (SOIREE) in 1999 detected by satellite imagery (ACE CRC 2008b, p. 5) – uncertainties remain over the efficacy, risk and impacts of such activities as large-scale climate change mitigation strategies (ACE CRC 2008b, p. 7).

Climate change has a number of related impacts. Warming temperatures from an increase in the levels of greenhouse gases emitted into the atmosphere heat the sea, with the consequent thermal expansion contributing to sea-level rise. At the same time, this warming reduces the ocean's ability to absorb carbon dioxide from the atmosphere as part of the natural carbon cycle. The uptake of carbon dioxide by the oceans leads to changes in ocean chemistry, altering its pH (the measure of acidity) to make the ocean more acidic. Given that the saturation horizon of CO_2 is temperature dependent, the impacts of acidification of the oceans are likely to be found first in the Southern Ocean. Recent research in the field has shown that one impact of changing ocean chemistry is the reduction in shell thickness of foraminifera, which are key microbial components in ocean ecosystems (ACE CRC 2008a). This may have considerable consequences for the marine ecosystem. Similarly, calcium carbonate structures such as coral reefs and coccolithophorids (Moy et al. 2009) are vulnerable to ocean acidification, which inhibits the formation of the calcium carbonate critical to their existence.

Warming is likely to have impacts on the cryosphere, including the significant sea-ice zone around Antarctica, and the Antarctic ice shelves and ice sheets. Research in the Southern Ocean has indicated changes in temperature (cooler waters) and salinity (reduced salinity of deep water compared with results from 10 years earlier) (Aoki et al. 2005a). This freshening of the oceans is likely to be caused by glacial melt from Antarctica and has the potential to affect the circulation around Antarctica (Aoki et al. 2005b). Changes in sea-ice extent and thickness have a number of potential impacts on ecosystems, and warming of oceans will increase subsurface melt of ice shelves. Ice shelves buttress glaciers and glacier discharge will accelerate when ice shelves are removed (ACE CRC 2009a). These processes and the changes likely to arise from them are identified in the scientific literature, but have considerable uncertainties (owing to the lack of data and understanding of the impacts and consequences arising from these changes due to the complexities of the system) associated with the rate and extent of warming-induced change on the Antarctic cryosphere.

2. CLIMATE CHANGE AND THE SOUTHERN OCEAN: REGIMES, INSTRUMENTS AND INSTITUTIONS

The Antarctic region is a major laboratory for research on climate change because its unique environment allows for equally unique scientific research to be undertaken (for example ice-core research on past climate, based on analysis of air trapped in ice initially frozen hundreds of thousands, perhaps up to a million years ago). For example, it has been noted that: '[t]he ice sheet provides data on past climate – for example, bubbles of gas trapped in ice reveal the extent to which human emissions have elevated greenhouse gas concentrations above natural levels' (Bergin and Haward 2007, p. 7). Research on ice cores provides long-term records of past climate; for example, the Law Dome core drilled by Australian scientists provides a record of 90 thousand years before the present. This paleoclimate research provides important data on the stability of ice sheets and ice shelves, and provides data on the composition of the atmosphere before the advent of anthropogenic emissions of greenhouse gases associated with the large-scale burning of fossil fuels. The Antarctic ice sheets are, as noted above, important in terms of the global climate, but 'a major unanswered question is the stability of the ice sheet itself. The 2007 IPCC *Fourth Assessment Report* acknowledges the potential for accelerating ice discharge by processes that are poorly understood' (ACE CRC 2009b).

2.1 The Antarctic Treaty System

The Antarctic Treaty was signed in 1959 by the 12 nations that had been active in Antarctica during the IGY. One key driver for the negotiation of the Antarctic Treaty was to allow the scientific cooperation enjoyed during the IGY to continue indefinitely, witnessed by Articles II and III, which provide for the freedom of scientific investigation and the promotion of international cooperation and free exchange of information and personnel. The Treaty also addressed and resolved the longstanding 'Antarctic Problem' with agreement over Article IV of the treaty, which sought to resolve issues relating to claims of territory and sovereignty on the continent (Christie 1951). Article IV has been characterized in a clichéd formulation as effectively 'freezing' existing and any further territorial claims to Antarctica, while allowing States to recognize claims. As Article IV of the Treaty states:

> 1. Nothing contained in the present Treaty shall be interpreted as
>   a) a renunciation by any Contracting Party of previously asserted rights of or claims to territorial sovereignty in Antarctica.
> 2. No acts or activities taking place while the present Treaty is in force shall constitute a basis for asserting, supporting or denying a claim to territorial sovereignty in Antarctica or create any rights of sovereignty in Antarctica. No new claim, or enlargement of an existing claim, to territorial sovereignty in Antarctica shall be asserted while the present Treaty is in force.

The Antarctic Treaty, which applies to the area south of 60° South Latitude, includes a commitment that Antarctica should not become the scene or object of international discord, a point reiterated in the text of the 1991 Protocol to the Antarctic Treaty (see below), which stipulates that Antarctica be devoted to peace and science. Here, again, science is linked to the Antarctic Treaty's focus as a peace treaty, overcoming tensions between national interests.

The Antarctic Treaty, in summary:

- stipulates that Antarctica should be used exclusively for peaceful purposes – military activities, such as the establishment of military bases or weapons testing, are specifically prohibited
- guarantees freedom to conduct scientific research
- promotes international scientific cooperation and requires that the results of research be made freely available
- sets aside the potential for sovereignty disputes between Treaty Parties by providing that no activities will enhance or diminish previously asserted positions with respect to territorial claims,

provides that no new or enlarged claims can be made, and makes rules relating to jurisdiction
- prohibits nuclear explosions and the disposal of radioactive waste in the Treaty area
- provides for inspection to ensure compliance with the Treaty – a world first weapons inspection system
- requires advance notice of expeditions
- provides for the Parties to discuss measures to further the Treaty (Bergin and Haward 2007).

The Antarctic Treaty now has 50 Parties, 28 of whom are Consultative Parties (ATCPs) on the basis of being original signatories or because they conduct substantial Antarctic research. Original signatories have decision-making status at Consultative Meetings by virtue of Article IX of the Treaty. Article XIII provides for accession of other States to the Treaty, and they too can achieve Consultative Party status by 'conducting substantial research activity [in Antarctica], such as the establishment of a scientific research station or the despatch of a scientific expedition' (Article IX.2).

Since entering into force in 1961, the Treaty has been recognized as one of the most successful international agreements. Differences over territorial claims have been effectively set aside, and as a disarmament regime the Treaty has been exceptionally successful (Haward et al. 2006). The focus on collaborative research developed through the IGY has continued to provide important contributions to earth systems science, and to global efforts to understand the processes and impacts of climate change (Jabour and Haward 2009).

This maintains the salience of science in Antarctic and Southern Ocean regimes, as science has been the 'currency' of the Antarctic Treaty regime (Herr and Hall 1989). Scientific advice is central to CCAMLR, and to the work of the CCAMLR Commission, which is charged with using an ecosystem approach to the management of Antarctic marine living resources. Science is similarly embedded within the Protocol on Environmental Protection to the Antarctic Treaty. Article 2 of the Protocol states that: '[t]he Parties commit themselves to the comprehensive protection of the Antarctic environment and dependent and associated ecosystems and hereby designate Antarctica as a natural reserve, devoted to peace and science'. The focus on environmental protection in the Protocol has encouraged new, multi-disciplinary science programmes among the Treaty Parties and reinforced international collaboration, as witnessed in the development of the 2007–2009 International Polar Year (IPY). As noted above, climate change has also become an important scientific driver for Antarctic research programmes. This is reflected in the addition of a climate change item in the recent agendas of the Antarctic Treaty Consultative Meetings (ATCM) (Jabour and Haward 2009).

2.2 The International Polar Year

The third IPY was held between 1 March 2007 and 1 March 2009 and involved 63 nations, 400 projects and approximately 50 000 scientists (Shadian and Tennberg 2009, p. 1). This is compared to the first IPY, which occurred in 1882–1883, and which saw 12 expeditions to the Arctic and three to Antarctica, involving 12 countries. The second IPY, established 50 years later in 1932–1933 involved 44 countries (Jabour and Haward 2009). Developments in science and technology following World War II encouraged the development of further international collaboration on polar science, with a focus on emergent areas of space research. This led to the establishment of the IGY in 1957–1958. The IGY was particularly influential in shaping Antarctic science and facilitating the collaboration later embodied in the 1959 Antarctic Treaty.

Threats to polar environments from human and environmental changes, such as those caused by climate change and technological developments, are leading to increasing challenges to governance in polar areas. The need to assess, quantify and understand these impacts, changes, challenges and threats was a major driver in the announcement of the third IPY. The IPY followed 50 years after the IGY and was to be two years in duration to ensure a full summer and winter of scientific research in both poles. Research undertaken in the IPY will be reported through 2012 and beyond (Jabour and Haward 2009).

The third IPY was established through the IPY Joint Committee of the International Council for Science (ICSU) and the World Meteorological Organization (WMO), co-chaired by Australia and Canada (ICSU 2004). This committee provides planning and oversight of the IPY, including designating core IPY projects.

The IPY focused on research on the atmosphere, ice, land, oceans, people and space, within six broad themes:

- Theme 1: To determine the present environmental status of the Polar Regions
- Theme 2: To quantify, and understand, past and present natural environmental and social change in the Polar Regions, and to improve projections of future change
- Theme 3: To advance our understanding on all scales of the links and interactions between Polar Regions and the rest of the globe, and of the processes controlling these links

- Theme 4: To investigate the frontiers of science in the Polar Regions
- Theme 5: To use the unique vantage point of the Polar Regions to develop and enhance observatories from the interior of the Earth to the Sun and the cosmos beyond
- Theme 6: To investigate the cultural, historical and social processes shaping the sustainability of circumpolar human societies, and to identify their unique contributions to global cultural diversity and citizenship (ICSU 2004, p. 7).

In an interesting development, the IPY has emphasized Education and Outreach, linking IPY science to education programmes as part of engendering interest in the Polar Regions by young people (the future polar scientists, managers and policy makers).

IPY 2007–2009 provided a significant impetus to Antarctic science, although there was some concern that the impact of the IPY's programmes on the broader population would not reach the heights of public interest that occurred with the earlier IGY as a result of the launch of the 'Sputnik' satellite (Walton 2003). In addition to terrestrial Antarctic programmes, a major marine science research program, the Census of Antarctic Marine Life (CAML) was undertaken (see CAML 2012 for more information). CAML involved major research cruises and multinational collaboration (Jabour and Haward 2009).

2.3 Law of the Sea Convention

Under the LOSC, coastal States have sovereignty over their territorial seas and sovereign rights, primarily for resource-related purposes, over their exclusive economic zones (EEZs). All States have broader freedoms on the high seas beyond national jurisdiction. High-seas freedoms have traditionally been seen as fundamental or unfettered rights, although these rights are constrained by a number of obligations and management arrangements that cover high-sea areas. Environmental protection is one such fundamental responsibility acknowledged by Parties to the LOSC. Sea-level rise, associated with global climate change, is likely to impact on the scope of claims to maritime jurisdiction as a consequence of the retreat of baselines from which maritime zones are measured (see the chapter by Richard Kenchington and Robin Warner in this volume).

The nexus between the LOSC and the Antarctic Treaty and its related instruments is increasingly important in managing the Southern Ocean (Haward 2009a). More recently, the intersection of the LOSC and the ATS has helped focus attention on questions of sovereignty (Vigni 2001). The declaration of maritime zones off the Australian Antarctic

Territory (AAT) by Treaty Parties has been controversial (e.g., see Kaye 1995, p. 191), but was in practice resolved by not enforcing territorial sea rights. The question of the definition of the limits of extended continental shelf areas off the AAT posed important political and legal challenges to Australia, but also a dilemma in relation to its interests. Australia had included the continental shelf of the AAT in its 1953 declaration of an Australian continental shelf, a move that pre-dated the negotiation and entry into force of the Antarctic Treaty and the LOSC. Failure to include the AAT in survey work and data collection for the extended continental shelf would arguably have indicated that Australia was weakening its claims to sovereignty over the AAT (Baird 2004; Serdy 2005). The solution to this dilemma was relatively simple. The Australian submission to the United Nations Commission on the Limits of the Continental Shelf (CLCS) included an outer limit for extended continental shelf areas off the AAT, but requested 'the Commission in accordance with its rules not to take any action for the time being with regard to the information in this submission that relates to continental shelf appurtenant to Antarctica' (The Permanent Mission of Australia 2004). Australia lodged the data for its claims to an extended continental shelf and requested that the CLCS not examine the data associated with the AAT (The Permanent Mission of Australia 2004). Australia's definition of the outer limits of its Antarctic continental shelf utilized a range of provisions available to Australia under Article 76 of the LOSC. This included the sediment thickness formula, and, as a result, provided significant data for further scientific research, including for climate reconstructions, as part of Australia's research on global climate change.

2.4 United Nations Framework Convention on Climate Change

The UNFCCC encourages all governments to achieve stabilization of greenhouse gas emissions at 1990 base year levels. Individual state sovereignty, economic but sustainable development, and reduction of emissions are guiding principles within the UNFCCC. It asks Parties to promote, among other things, enhancement of natural sinks and reservoirs, including the oceans and marine ecosystems, where appropriate. The UNFCCC provides for annexes and protocols to be attached to the parent document to supply further detail as information becomes available. The Kyoto Protocol on greenhouse gas emissions was adopted in Kyoto, Japan, on 11 December 1997 and entered into force on 16 February 2005:

> The Kyoto Protocol is an international agreement linked to the United Nations Framework Convention on Climate Change. The major feature of the Kyoto

> Protocol is that it sets binding targets for 37 industrialized countries and the European community for reducing greenhouse gas (GHG) emissions. These amount to an average of five per cent against 1990 levels over the five-year period 2008–2012. (UNFCCC 2009)

The 1997 Kyoto Protocol to the UNFCCC, while not specifically mentioning oceans, asks the Parties to protect and enhance carbon sinks and reservoirs, and to research, promote, develop and increase the use of sequestration technologies.

2.5 The Intergovernmental Panel on Climate Change

The Intergovernmental Panel on Climate Change (IPCC), established by the United Nations Environment Programme and the WMO, has been charged with providing expert assessments of climate change (IPCC 2009).

> The IPCC is a scientific body. It reviews and assesses the most recent scientific, technical and socio-economic information produced worldwide relevant to the understanding of climate change. It does not conduct any research nor does it monitor climate related data or parameters. Thousands of scientists from all over the world contribute to the work of the IPCC on a voluntary basis. Review is an essential part of the IPCC process, to ensure an objective and complete assessment of current information. (IPCC 2009)

The IPCC has completed four assessment reports in 1990, 1995, 2001 and 2007. Work on the fifth assessment report, due in 2014, has begun. Since the IPCC does not commission research, but merely collates and reviews peer-reviewed (published) science, the work of national Antarctic programmes and international collaborative research efforts are increasingly important in addressing significant uncertainties, including those related to sea-level rise. The IPCC's *Fourth Assessment Report (*AR4), published in 2007, estimated the maximum projected total sea-level rise by 2090–2099, excluding any accelerated ice discharge, as 59 cm (ACE CRC 2009b). AR4 considered that a further 20 cm of sea-level rise might occur due to rapid dynamic change to the ice sheets, but noted that this is an area of significant uncertainty (ACE CRC 2009b) (for more information see the chapter by in Richard Kenchington and Robin Warner in this volume).

2.6 Convention on Biological Diversity

The CBD entered into force in December 1993. Each Party to the Convention has responsibility for the conservation and sustainable use of its own biological diversity, and Parties are to cooperate in implementing the Convention in areas beyond national jurisdiction such as the high seas (Haward and Vince 2008, p. 46). The Jakarta Mandate, negotiated at the second meeting of States Parties to the CBD in 1995, centred on the application of the CBD to marine and coastal environments (Haward and Vince 2008). The intersection between the CBD and the ATS has been limited, although the former has been promoted as a means to support efforts to manage biological prospecting (bio-prospecting) of marine genetic and biochemical resources (Jabour and Nicol 2003). While such initiatives have not been supported thus far, concern over the impact of climate change on Southern Ocean biological diversity is likely to enhance interest in addressing these issues, and in the region, through the CBD processes.

The ninth and tenth Conference of Parties (COP 9 Bonn, Germany 19–30 May 2008 and COP 10 Nagoya, Japan 18–29 October 2010) to the CBD discussed ocean fertilization as part of its agenda item on biodiversity and climate change. A resolution was agreed upon that, among other matters, requested Parties and urged governments, in accordance with the precautionary approach, to ensure that ocean fertilization activities do not take place until there is an adequate scientific basis on which to justify such activities and to assess associated risks (Djoghlaf 2008). Some Parties to the CBD opposed the resolution, or attempts at stronger measures within the framework of the Convention, as they believed that the issue was being addressed in other forums, and the CBD action was precipitate and counter to the work being undertaken. According to the CBD resolution, with the exception of small-scale scientific research studies within coastal waters, there should be a global, transparent and effective control and regulatory mechanism in place for ocean fertilization activities (CBD COP 9 2008 Decision IX/16 C).

CBD COP 10 recorded, 'there is a pressing need for research to advance our understanding of marine ecosystem dynamics and the role of the ocean in the global carbon cycle' (CBD COP 10 2010 Decision X29 62) and noted the ongoing work on ocean fertilization, including that undertaken by the London Convention and Protocol. Decision X33 of COP 10 included that States will:

> ensure, in line and consistent with decision IX/16 C on ocean fertilization and biodiversity and climate change, in the absence of science-based, global, transparent and effective control and regulatory mechanisms for geo-engineering, and in accordance with the precautionary approach and Article 14 of the Convention, that no climate-related geo-engineering activities that may affect biodiversity take place, until there is an adequate scientific basis on which to justify such activities and appropriate consideration of the associated risks for the environment and biodiversity and associated social, economic and cultural impacts, with the exception of small-scale scientific research studies that would be conducted in a controlled setting in

> accordance with Article 3 of the Convention, and only if they are justified by the need to gather specific scientific data and are subject to a thorough prior assessment of the potential impacts on the environment. (CBD COP 10 2010 Decision X33 8w)

Decision IX/16 C CBD COP 9 (reiterated in COP 10 Decision X33 8w) was significant and interpreted by some parties as a moratorium or ban on all open ocean fertilization, including for research. In June 2008, the Intergovernmental Oceanographic Commission (IOC) Ad Hoc Consultative Group on Ocean Fertilization expressed concern over this decision, on the basis that it would impede legitimate research activities. Meetings of Contracting Parties to the London Convention and London Protocol have also addressed the CBD COP 9 decision.

2.7 Convention/Protocol on the Prevention of Marine Pollution by Dumping of Wastes and Other Matter (London Convention/London Protocol)

The Convention/Protocol on the Prevention of Marine Pollution by Dumping of Wastes and Other Matter (London Convention/London Protocol) provides a regulatory framework for dumping of wastes at sea. Following initial discussions at the 1972 United Nations Conference on the Human Environment in Stockholm, an intergovernmental conference in London concluded the Treaty in late 1972 (Birnie and Boyle 1995). The Convention entered into force in 1975 and in 1996 a Protocol was developed to modernize the Convention 'and, eventually, replace it' (IMO 2009).

There are international legal restrictions on what can be put into the ocean under the 1996 Protocol to the London Convention. All intentional dumping is prohibited, except for possibly acceptable wastes (the London Protocol uses a reverse list procedure *permitting* specific substances to be dumped into the ocean and *prohibiting* those not listed). A 2006 amendment to the Protocol has added CO_2 streams from CO_2 capture processes to the list of permitted substances. This amendment indicates a broadening of the scope of the Protocol in relation to greenhouse gas mitigation processes. The amendment, which entered into force on 10 February 2007, states that a CO_2 stream may only be considered for disposal if:

- it is into a sub-seabed geological formation
- it consists overwhelmingly of CO_2 (it may contain incidental associated substances derived from the source material and the capture and sequestration processes used)
- no wastes or other matter are added for the purpose of disposing of them.

3. CLIMATE CHANGE AND THE SOUTHERN OCEAN: SCIENCE, LAW AND POLICY – THE CASE OF OCEAN FERTILIZATION

Ocean fertilization experiments that have been undertaken thus far (10 in total) have been classed as marine scientific research, sanctioned through Part XIII of the LOSC. With the exception of LOHAFEX, undertaken in 2009 after the passing of the CBD resolution, neither the scale nor the type of experiment had caused any controversy or complications. However, the shift from scientific research to commercial activity raises a number of issues. For example, if the locations are in areas outside national jurisdiction, benefit-sharing regimes under the CBD and, to a lesser extent, those under Part XI of the LOSC could be invoked. Article 140 of the LOSC notes that 'Activities in the Area shall, as specifically provided for in this Part, be carried out for the benefit of mankind as a whole'. Article 140 further states that benefits to mankind as a whole are 'irrespective of the geographical location of States' or whether they are coastal or land-locked. Such activities shall also take into account the interests and needs of developing States and peoples who have not gained independence or self-governing status. Part 2 of Article 140 mandates equitable sharing of financial and other economic benefits. Commercial activity may also trigger different issues under these and other international instruments including the Antarctic Treaty/Protocol and the London Convention/London Protocol. One of the most significant matters for concern will be that of harm to the marine environment.

The London Convention/Protocol provides a general obligation on all Parties to act in a precautionary manner in the absence of scientific *certainty* and to do nothing that could cause trans-boundary environmental harm. Under the London Protocol, the Parties:

> shall apply a precautionary approach to environmental protection from dumping of wastes or other matter whereby appropriate preventative measures are taken when there is reason to believe that wastes or other matter introduced into the marine environment are likely to cause harm even when there is no conclusive evidence to prove a causal relation between inputs and their effects. In implementing the provisions of this Protocol, Contracting Parties shall act so as not to transfer, directly or indirectly, damage or likelihood of damage from one part of the environment to another or transform one type of pollution into another. (London Protocol 1996, arts 3.1 and 3.3)

As noted above, scientific research on ocean fertilization has identified the potential for increased CO_2 uptake from added iron or similar elements to the ocean to stimulate phytoplankton growth. At the same time, the impacts of such manipulation, if undertaken on the scale needed to enhance oceanic drawdown of CO_2 as a climate change mitigation strategy, are unknown (Broder and Haward in press). Concerns over unregulated ocean fertilization activities have been raised at international forums. On 13 July 2007, the London Convention Protocol (LC/LP) Scientific Groups released a Statement of Concern on ocean fertilization, noting that while iron fertilization may offer a potential strategy for removing carbon dioxide, the process is still only speculative (IMO 2007).

Following increased attention on ocean fertilization, States Parties to the London Convention/Protocol continued their work on a regulatory framework. The October 2008 meeting of Parties to the London Convention and Protocol (Meetings number 30 and 3, respectively) passed Resolution LC/LP 1, which, among other points, limits ocean fertilization to activities for 'legitimate scientific experimentation'. The LC/LP Scientific Groups were requested to draft an assessment framework to help parties determine appropriate activities (Broder and Haward in press). Following the completion of this framework, in 2010, at meetings number 32 and 5, respectively, Parties to the London Convention/Protocol adopted the risk assessment framework (Broder and Haward in press).

The question of competency over regulation of iron fertilization activities was given impetus by the almost contemporaneous discussions within the IMO supported Scientific Groups to the London Convention/London Protocol, and the CBD in May 2008. While these deliberations resulted in different outcomes and potentially conflicting obligations for States Parties to both the CBD and the London Convention/London Protocol, ongoing work has narrowed this gap. As noted above, the CBD's most recent COP referenced the work of the London Convention/Protocol, but noted that any ocean fertilization/nourishment activities have to be 'justified by the need to gather specific scientific data and [be] subject to a thorough prior assessment of the potential impacts on the environment'. The development of a risk-based assessment framework by the Scientific Groups to the London Convention/London Protocol reinforces the intersection between science, law and policy related to climate change and Antarctica and the Southern Ocean.

4. CONCLUSION

This chapter has argued that the Southern Ocean and Antarctica have a critical role in global and regional climate through the interaction of elements of the cryosphere, including wind, ice and ocean. The Southern Ocean plays a major role in global climate, but is also a region of increasing resource exploitation, and a region in which a number of regimes, and therefore science, law and policy, intersect. Science has been a driver of Antarctic endeavour for a century, and is embedded in the Antarctic Treaty. These efforts are increasingly important in addressing major uncertainties related to climate change. In addition, the topic and the underlying science are being considered by a number of forums. The question of regime intersection or interplay also raises interesting theoretical and analytical issues: a key question being, 'how can we diagnose that international agreements interact with each other, and how do we separate and aggregate the effects of multiple regulatory instruments?' (Sprinz 2007, p. 1).

As noted above, the right to use the world's oceans is a fundamental freedom enshrined in the LOSC for coastal States within their territorial sea and EEZs. Marine scientific research is further protected, and is central to work to increase understandings of the rate and extent of climate change and variability. Research activities in Antarctica and the Southern Ocean, supported under the ATS regime, will also be influenced by a number of other international instruments, depending on the location and purpose of the proposed activity. This leads to a number of practical issues related to overlap and gaps in regimes, questions of competency (i.e., which instrument or institution should manage the issue or problem) and the potential consequences of 'forum shopping' – the process whereby States may select instruments or institutions to deal with an issue on the basis that they are more likely to provide a desirable outcome. Climate change is clearly a problem area that lends itself to this type of analysis, resulting from the intersection between science, law and policy, with each being subject to a range of instruments and institutional dynamics. Examination of these interactions (forming the basis of the concept of 'climate services', see World Climate Conference 3 2009) in Antarctica and the Southern Ocean highlights the need to consider carefully any consequences related to the question of overlap (and perceived gaps) between regimes.

NOTE

* The author would like to thank colleagues at the Antarctic Climate and Ecosystems Cooperative Research Centre, University of Tasmania for providing comments, information and an understanding of the complex interactions underpinning climate change science. The author remains responsible for all errors of omission or commission. The author also acknowledges the support of the Antarctic Climate and Ecosystems Cooperative Research Centre, funded through the Australian Government's Cooperative Research Centre Programme.

REFERENCES

AAD (2002), *Looking South: The Australian Antarctic Program into the 21st Century*, Kingston: AAD.

ACE CRC (2008a), *Position Analysis: CO2 Emissions and Climate Change: Ocean Impacts and Adaptation Issues*, Hobart: ACE CRC.

ACE CRC (2008b), *Position Analysis: Ocean Fertilisation: Science and Policy Issues*, Hobart: ACE CRC.

ACE CRC (2009a), *Position Analysis: Polar Ice Sheet and Climate Change: Global Impacts*, Hobart: ACE CRC.

ACE CRC (2009b), *Position Analysis: Changes to Antarctic Sea Ice: Impacts*, Hobart: ACE CRC.

Aoki, S., S.R. Rintoul, S. Ushio, S. Watanabe and N.L. Bindoff (2005a), 'Freshening of the Adélie Land Bottom Water near 140°E', *Geophysical Research Letters*, **32**, L23601.

Aoki, S., N. Bindoff and J.A. Church (2005b), 'Interdecadal water mass changes in the Southern Ocean between 30°E and 160°E', *Geophysical Research Letters*, **32**, L07607.

Argo (2010), 'Origins of Argo', accessed 15 February 2010 at: www.argo.ucsd.edu/Origins_of_Argo.html.

Baird, R. (2004), 'Can Australia assert an extended continental shelf off the Australian Antarctic Territory consistent with the Law of the Sea and within the constraints of the Antarctic Treaty?' *Maritime Studies*, **138**, 1–19.

Bergin, A. and M. Haward (2007), *Frozen Assets: Securing Australia's Antarctic Future*, Strategic Insight No 34, Canberra: Australian Strategic Policy Institute.

Birnie, P.W. and A. Boyle (1995), *Basic Documents on International Environmental Law*, Oxford: Clarendon Press.

Bork, K., J. Karstensen, M. Visbeck and A. Zimmerman (2008), 'The legal regulation of floats and gliders – in quest of a new regime?' *Ocean Development and International Law*, **39** (3), 298–328.

Broder, S. and M. Haward (in press) 'The international legal regimes governing ocean iron fertilization', in J.-H. Paik, S. Lee and H. Scheiber (eds), *Institutions and Regions in Ocean Governance*, Leiden: Martinus Nijhoff.

CAML (2012), *Census of Antarctic Marine Life*, accessed 26th July 2012 at: http://www.caml.aq/.

Christie, E.W.H. (1951), *The Antarctic Problem: A Historical and Political Study*, London: Allen and Unwin.

Convention on Biological Diversity, Ninth Meeting of the Conference of Parties (CBD COP 9) Decision IX/16 Biodiversity and Climate Change, accessed 26 July 2012 at: http://www.cbd.int/decision/cop/?id=11659.

Convention on Biological Diversity, Tenth Meeting of the Conference of Parties (CBD COP 10), Decision X 29 Marine and Coastal Biodiversity, accessed 26 July 2012 at: http://www.cbd.int/decision/cop/?id=12295.

Convention on Biological Diversity, Tenth Meeting of the Conference of Parties (CBD COP 10), Decision X33 Biodiversity and Climate Change, accessed 26 July 2012 at: http://www.cbd.int/decision/cop/?id=12299.

Djoghlaf, A. (2008), 'Message of Dr Ahmed Djoghlaf, Executive Secretary, on the Occasion of World Ocean Day, 8 June 2008, Secretariat of the Convention on

Biological Diversity', accessed 26th July 2012 at: www.cbd.int/doc/speech/2008/sp-2008-06-08-ocean-en.pdf.
Haward, M. (2009a), *Environmental Security in the Southern Ocean: An Ongoing Agenda?* Scoping Paper for Workshop Responding to Contemporary Challenges and Threats to Antarctic Security: Legal and Policy Perspectives, 5–7 July 2009, School of Law and Gateway Antarctica, University of Canterbury, Christchurch, New Zealand.
Haward, M. (2009b), 'The Law of the Sea Convention and the Antarctic Treaty System: constraints or complementarity', in S.-Y. Hong and J. van Dyke (eds), *Maritime Boundary Disputes, Settlement Processes, and the Law of the Sea*, Leiden: Martinus Nijhoff Publishers, pp. 231–251.
Haward, M., D.R. Rothwell, J. Jabour, R. Hall, A. Kellow, L. Kriwoken, G. Lugten, and A. Hemmings (2006), 'Australia's Antarctic agenda', *Australian Journal of International Affairs*, **60** (3), 439–456.
Haward, M. and J. Vince (2008), *Oceans Governance in the Twenty-First Century: Managing the Blue Planet*, Cheltenham, UK and Northampton, MA: Edward Elgar.
Herr, R.A. and H.R. Hall (1989), 'Science as currency and the currency of science', in J. Handmer (ed.), *Antarctica: Policies and Policy Development*, CRES Resource and Environmental Studies No 1, Canberra: CRES, ANU.ICSU (2004), *A Framework for the International Polar Year 2007–2008*, ICSU IPY 2007–2008 Planning Group.
International Maritime Organization (IMO) (2007), Convention on the Prevention of Marine Pollution by Dumping of Wastes and Other Materials, 1972 and its 1996 Protocol. Statement of Concern Regarding Iron Fertilization of the Ocean to Sequester CO_2. LC-LP. 1/Circ. 14, 13 July 2007.
IMO (2009), 'London Convention 1972', accessed 25 July 2012 at: http://www.imo.org/OurWork/Environment/SpecialProgrammesAndInitiatives/Pages/London-Convention-and-Protocol.aspx.
IPCC (2009), IPCC Organization, last accessed 13 July 2009 at: www.ipcc.ch/organization/organization.htm.
Jabour J. and M. Haward (2009), 'Antarctic science, politics and IPY legacies' in J.M. Shadian and M. Tennberg (eds), *Legacies and Change in Polar Sciences: Historical, Legal and Political Reflections on the International Polar Year*, Farnham: Ashgate, pp. 101–124.
Jabour, J.A. and D. Nicol (2003), 'Bioprospecting in areas outside national jurisdiction: Antarctica and the Southern Oceans', *Melbourne Journal of International Law*, **4** (1), 76–111.
Kaye, S. (1995), *Australia's Maritime Boundaries*, Wollongong Papers in Marine Affairs No 4, Wollongong: Centre for Maritime Policy, University of Wollongong.
Moy, A.D., W. Howard, S.G. Bray and T. Trull (2009), 'Reduced calcification in modern Southern Ocean planktonic foraminifera', *Nature Geoscience*, **2**, 276–280.
London Protocol (1996), Protocol to the Convention on the Prevention of Marine Pollution by Dumping of Wastes and Other Materials, London, 7 November 1996 36 I.L.M. 1.
Rahmstorf, S. (2006), 'Thermohaline ocean circulation', in S.A. Elias (ed.), *Encyclopedia of Quaternary Sciences*, Amsterdam: Elsevier.
Rintoul, S. (2010), CSIRO and ACE CRC, personal communication, 17 February.

Serdy A. (2005), 'Towards certainty of seabed jurisdiction beyond 200 nautical miles from the territorial sea baseline: Australia's submission to the commission on the limits of the continental shelf', *Ocean Development and International Law*, **36**, 201–217.

Shadian, J. and M. Tennberg (2009), 'Introduction', in J.M. Shadian and M. Tennberg (eds), *Legacies and Change in Polar Sciences: Historical, Legal and Political Reflections on the International Polar Year*, Farnham: Ashgate, pp. 1–6.

Sprinz, D. (2007), *Regime Effectiveness: The Next Wave of Research*, paper presented at 48th Annual Convention of the International Studies Association, 28 February–3 March 2007, Hilton, Chicago.

Steffen, W. (2006), *Stronger Evidence but New Challenges: Climate Change Science 2001–2005*, Canberra: Department of Environment and Heritage, Australian Greenhouse Office.

Steig, E.J., D.P. Schneider, S.D. Rutherford, M.E. Mann, J.C. Comiso and D.T. Shindell (2009), 'Warming of the Antarctic ice-sheet surface since the 1957 International Geophysical Year', *Nature*, **457**, 459–462.

Stoddart, M. and M. Haward (2011), 'Science', in M. Haward and T. Griffiths (eds), *Australia and the Antarctic Treaty System: Fifty Years of Influence*, Sydney: UNSW Press. pp. 138–163.

The Permanent Mission of Australia to the United Nations (2004), *Note from the Permanent Mission of Australia to the Secretary-General of the United Nations Accompanying Lodgment of Australia's Submission*, Note No 89/2004, accessed 26th July 2012 at: http://www.un.org/Depts/los/clcs_new/submissions_files/aus04/Documents/aus_doc_es_attachment.pdf.

Turner, J.T.A. Lachlan-Cope, S. Colwell, G.J. Marshall and W.M. Connolley (2006), 'Significant warming of the Antarctic winter troposphere', *Science*, **311**, 1914–1917.

UNFCCC (2009), 'UNFCCC', accessed 25 July 2012 at: http://unfccc.int/kyoto_protocol/items/2830.php.

Vigni, P. (2001), 'Antarctic maritime claims: "frozen sovereignty" and the Law of the Sea', in A.G. Oude Elferink and D.R. Rothwell (eds), *The Law of the Sea and Polar Maritime Delimitation and Jurisdiction*, The Hague: Martinus Nijhoff Publishers, pp. 85–104.

Walton, D. (2003), 'Editorial: The International Polar Year in 2007', *A*ntarctic Science, 15 (4), 413.

World Climate Conference 3 (2009), *Conference Statement*, World Climate Conference 3, Geneva, 31 August–4 September 2009.

6. Climate change and the limits of maritime jurisdiction

Clive Schofield and Andi Arsana

1. INTRODUCTION

Among the most salient impacts on the oceans associated with climate change is potentially significant and, critically, rapid sea-level rise. Sea-level rise of this character would clearly pose potentially disastrous threats to numerous coastal States, especially those with large areas of heavily-populated, low-lying coastal territory. In light of long-standing and ongoing shifts in the concentration of global populations from rural to urban and from interior/highland to coastal/lowland contexts, concerns over the potential effects of sea-level rise on heavily inhabited areas are significant (Schofield 2011). Indeed, it has been estimated that a sea-level rise of 1 m would inundate the territory presently home to around 60 million people (Ananthaswamy 2009). Further, even if not permanently inundated, climate change and sea-level rise are likely to make low-lying coastal areas more vulnerable to periodic flooding as a consequence of an increasing incidence of extreme weather events, leading, for instance, to storm surges occurring on top of an elevated base sea level (Gornitz 1995). Of particular concern in this context are large portions of the low-lying mega-deltas of the world, including those of Vietnam's Mekong and Red Rivers in the Asia-Pacific. These areas are especially vulnerable because large areas are actually below mean sea level at present sea levels (Doyle et al. 2010). Indeed, it has been estimated that a 1 m rise in sea level could displace more than 7 million inhabitants in the Mekong delta area alone (UNDP 2011). Moreover, in addition to threats to populated coastal areas, concerns have been raised over the potential impacts of increased salt water intrusion on agricultural land close to the coast as well as valuable coastal environments and habitats such as wetlands and mangroves as a consequence of sea-level rise (Freestone 1991). Rising sea levels and the inundation of coastal territory also pose a dire and possibly terminal threat for communities inhabiting low-lying islands and States composed of islands.

While the impact of sea-level rise on land territory and coastal populations and infrastructure has, understandably, been the most immediate and pressing concern for policymakers, sea-level rise also has significant potential to affect the extent of a coastal State's maritime jurisdiction as well as its land territory. This is a direct result of the fundamental link between national claims to maritime jurisdiction and baselines constructed along the coast. The critical issue in this context is that the predominant type of baselines from which maritime claims are measured globally – that is, normal baselines coinciding with the low-water line along the coast – are susceptible to change. Consequently, such baselines and the limits of maritime claims measured from them are vulnerable to sea-level rise. In short, as the sea level rises the location of normal baselines will tend to recede inland. As a result, the limits to national maritime jurisdictional zones may be correspondingly pulled back, leading to a significant diminution in the coastal State's maritime spaces. This necessarily has implications with respect to rights over and access to the valuable marine resources contained within the maritime areas lost to the coastal States concerned.

This chapter focuses on the potential effects of sea-level rise on coastal State claims to maritime jurisdiction. First, a brief overview of some key issues and uncertainties associated with global sea-level rise is provided. The critical link between baselines and limits of zones of maritime jurisdiction is then explored. The way baselines and therefore maritime limits have the capacity to move over time in accordance with the dynamic nature of coasts, together with potential impacts of sea-level rise in this context, is illustrated. The probable uneven character of such impacts is highlighted with reference to especially vulnerable coasts, and some of the particular challenges faced in island contexts are also discussed. The chapter then addresses some of the options available to States in response to sea-level rise – in particular the threat it poses to baselines and maritime claims. There is a specific focus on ways the limits of maritime jurisdiction are now determined, and how currently vulnerable or impermanent maritime limits may be fixed in the future. Concluding thoughts are then offered.

2. CLIMATE CHANGE AND SEA-LEVEL RISE

Sea-level rise is an inevitable consequence of ocean warming due to the thermal expansion of the oceans, linked to the increasing atmospheric temperatures associated with global warming. As air temperatures rise, so the oceans warm gradually and incrementally. The warming of the ocean's surface waters leads them to expand, and this, in turn, translates to a rise in sea level (Gornitz 1995; Church et al. 2008). The other major cause of sea-level rise suggested is the melting of glaciers and grounded ice sheets (as opposed to those floating on the ocean).

There is broad agreement among the scientific community as to the reality of rising sea level on a global scale. For example, reviews of multiple studies of the issue suggest that global sea level has been rising at a rate of 1–2 mm per year over the last century, with an indication that the correct figure is closer to 2 mm per year (Douglas 1997). This has led to the observation that the contemporary rate of global sea-level rise 'stands in sharp contrast to the rate during the past several millennia' (Douglas 1997). It has also been suggested that this difference is indicative of 'a relatively recent climate signal' (Gornitz 1995). Concerns have also been expressed that the rate of increase in global sea level is accelerating (with a rate of rise of approximately 3.1 mm per year based on satellite measurements from the 1990s) (Church et al. 2008; Woodroffe 2011) and that the sea level will rise at a faster rate than currently as a consequence of global warming (Douglas 1997).

Despite substantial agreement within the scientific community as to the reality of sea-level rise on a global scale, considerable uncertainty persists concerning the causes of the phenomenon and the critical issues of the degree to which the sea will rise and how swiftly it will do so.

In large part, these uncertainties stem from the multitude of complex factors that may contribute to sea-level rise (and fall) and the interplay between them (Schofield 2011). For example, although the melting of glaciers and other grounded ice are well known and potentially very significant contributors to sea-level rise, the extent and speed of their melting remains highly debatable. For example, the Greenland ice sheet alone has enough water locked in it to raise global sea level by 6–7 m were it to collapse and melt completely, while the melting of the West Antarctic ice sheet would result in an estimated sea-level rise of 5 m (Walker and King 2008). While acknowledging this uncertainty, Walker and King nonetheless conclude that 'Greenland is one of the most convincing reasons we have for the urgent need to curb climate change' (Walker and King 2008, p. 80; see also, Hansen 2006). Similarly, although the calamitous possibility of the West Antarctic ice sheet disintegrating and melting leading to catastrophic sea-level rise was first suggested over three decades ago (Mercer 1968, 1978; see also Oppenheimer and Alley 2004), the apparent stability of this ice sheet (based on satellite observations from the late 1970s) has suggested that fears concerning the loss of such enormous volumes of ice were ill-founded (Gornitz 1995). That said, even limited melting of such huge ice sheets as a result of climate change would have profound impacts on global sea level. For example, it has been observed that the melting of the land-based ice in Antarctica's Amundsen Gulf alone could contribute approximately 1.5 m to global sea-level rise (Vaughan 2008).

As a consequence of the considerable uncertainty that surrounds the issues of whether and how swiftly land-based ice sheets such as those of Antarctica and Greenland are melting, the Intergovernmental Panel on Climate Change (IPCC) did not factor in this possible loss of ice. This led to its relatively moderate predictions in the *IPCC Fourth Assessment Report* of 2007. Consequently, the IPCC's estimates of the range of sea-level rise by 2100 of between 0.18 and 0.59 m above 1990 levels, with a mid-range prediction of 40 cm (IPCC 2007), have been criticized as overly conservative (see for example Ananthaswamy 2009).

It has also been recognized that sea-level rise is a phenomenon that exhibits marked spatial and temporal variability. Sea level varies diurnally, under the influence of the tides, but also seasonally, regionally and inter-annually. For example, complex atmospheric-oceanic interactions such as the El Niño Southern Oscillation (ENSO), can cause sea-level rise of up to 600 mm over periods of months every 5–10 years. Similarly, the Northern Atlantic Oscillation is responsible for variations in atmospheric pressure ranging from –700 to +1300 mm over hours to months (Gornitz 1995) (see Chapter 2). Sea level is also influenced by vertical displacements associated with tectonic movements, postglacial uplift (isostatic rebound), subsidence and anthropogenic impacts (Gornitz 1995; see also, for example Sutanta et al. 2009 and Alam 1996). Taken together, these factors make accurate sea level observation, let alone prediction, extremely challenging. It is also relevant to acknowledge that challenges exist in determining the extent of actual sea-level rise against the backdrop of ongoing long-term cyclical changes. All of these factors combine to influence sea-level rise (and fall) in variable ways, quite apart from the influences of climate change (Gornitz 1995). This led the Intergovernmental Panel on Climate Change (IPCC) to conclude in its Fourth Assessment Report of 2007 that sea level change is 'highly non-uniform spatially', noting that in some regions rates of rise are several times higher than the global mean rise while in other regions, sea level is falling (IPCC 2007). Moreover, the impacts on particular coastlines are substantially dependent on its particular characteristics, such as the morphology of the seabed immediately offshore (see below).

Nonetheless, despite continuing uncertainty over the critical questions of how much sea level will rise and how swiftly, coastal States are bracing for considerably higher sea level. For example all tide gauges around Australia indicate a rising trend, with tide gauges in northern Australia indicating rates of rise of more than two times the global average (Church et al. 2006). Accordingly, the Australian Government's Department of Climate Change published *Climate Change Risks to Australia's Coast: A First Pass National Assessment* in November 2009, which suggests a 'plausible worst-case scenario' of a rise of 1.1 m by 2100 (Australian

Government, Department of Climate Change 2009). It is anticipated that such a rise would result in inundation of or saltwater intrusion into environmentally sensitive and valuable low-lying areas such as the wetlands of the Kakadu National Park in the Northern Territory. Indeed it has been suggested that around 100 km^2 of wetlands (mainly mangroves) may be lost from Kakadu National Park because of inundation by 2100 (Australian Government, Department of Climate Change 2009). Further, in the order of 150 000–250 000 homes with an estimated value of around A$63 billion (approximately US$65.5 billion) could be at risk should a rise of this magnitude occur (Australian Government, Department of Climate Change 2009).

While it is appropriate to acknowledge that significant challenges exist in distinguishing sea-level rise from background variations or 'static' – that is, ongoing long-term cyclical changes as well as the influence of other processes – even the relatively modest sea-level rise envisaged by the IPCC would have severe consequences and pose major challenges for coastal States (Schofield 2011). The critical point here is that even if dire predictions of rapid multi-metre sea-level rise prove unfounded, even moderate vertical sea-level rise can have significant consequences horizontally where low-lying shallow gradient coastlines are concerned, and thereby potentially have significant consequences for the limits and extent of national claims to maritime jurisdiction.

3. BASELINES AND ZONES OF MARITIME JURISDICTION

A key advance made through the *United Nations Convention on the Law of the Sea* (LOSC) (United Nations 1983) was the definition of clear spatial limits for national claims to maritime jurisdiction. This represented an especially notable achievement given that it had not been achieved through earlier codification efforts. Measured seawards from its baselines, a coastal State is entitled to claim a series of zones of maritime jurisdiction provided for in accordance with the LOSC. These zones include a territorial sea of 12 nautical miles (nm) (LOSC, Article 3), a contiguous zone out to 24 nm from baselines (or 12 nm from territorial sea limits) (LOSC, Article 33), an exclusive economic zone (EEZ) out to 200 nm from baselines (LOSC, Article 57) and continental shelf rights that may extend up to 350 nm or even further in certain circumstances (LOSC, Article 76) (see Figure 6.1). The outer limits of the continental shelf where it extends beyond the 200 nm exclusive economic zone (EEZ) limits, unlike the limits of other zones that are based solely on a distance measurement, depend also on the

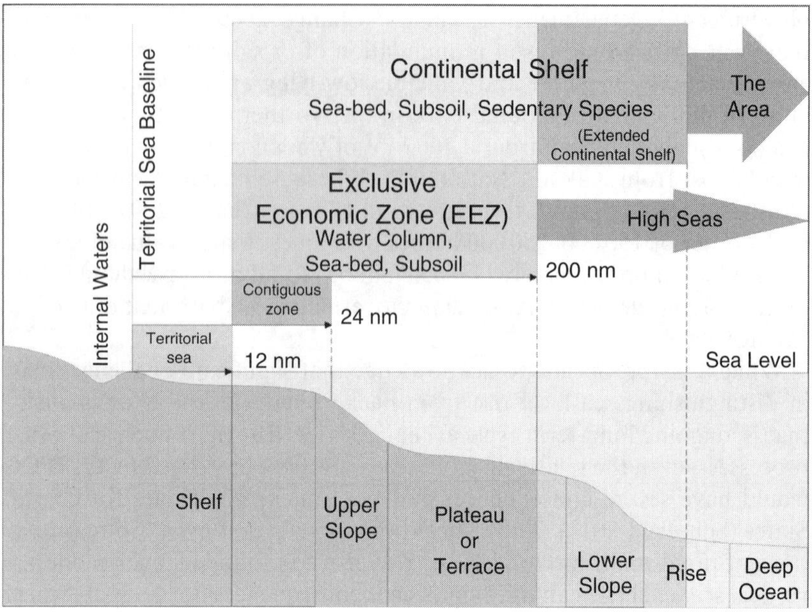

Figure 6.1 Baselines and zones of maritime jurisdiction

geology and geomorphology of the seabed. That is why determining the outer limit of the continental shelf is a more complex task than that for other zones of maritime jurisdiction. In addition to the aforementioned zones, a coastal State is also entitled to internal waters measured landward from baselines when applicable.

With respect to what are often termed 'territorial sea baselines', but are in fact the baselines also applicable to maritime zones other than the territorial sea, the LOSC allows for several different options. The default situation in the absence of any other baselines claim is that 'normal' baselines in accordance with Article 5 of the LOSC will apply. Alternatively and in specific geographical situations, a coastal State may designate straight baselines (LOSC, Article 7), lines closing the mouths of rivers (LOSC, Article 9) and the mouths of bays (LOSC, Article 10) as well as baselines related to the outermost permanent harbour works of ports (LOSC, Article 11). Further, a State that qualifies as an archipelagic State according to Article 46 of the LOSC can designate archipelagic baselines 'joining the outermost points of the outermost islands and drying reefs of the archipelago' (LOSC, Article 47). A State may use a combination of different types of baselines to construct the overall baseline around its coastline (LOSC, Article 16). It is, however, important to note that the various straight-line type baselines outlined above are all still linked to normal baselines, as they require anchoring turning and terminal points located on normal, low-water line baselines.

Irrespective of the type, baselines are vital in defining the outer limits of maritime zones claimed by a coastal State. Landwards of a coastal State's baselines lie either its land territory, including the inter-tidal foreshore landwards of normal low-water line baselines, or internal waters. Crucially, baselines also serve as the starting point from which the outer limits of maritime zones are measured. In addition, baselines are also critical to the construction of equidistance lines between coastal States in the delimitation of maritime boundaries. In this context it is notable that equidistance lines, the construction of which necessarily depend on the use of baselines, have proved to be the most popular method of delimitation by far (Prescott and Schofield 2005). Further, the approach adopted by international courts and tribunals in recent international cases relating to the delimitation of maritime boundaries has been to define a provisional boundary line based on equidistance, and then to examine any factors that may justify a modification of the provisional line in order to achieve an equitable result (Prescott and Schofield, 2005).

In their designation, normal baselines do not require coastal States to make any active claim or publication. Additionally Article 5 of LOSC does not specify a particular low-water line that should be used; this choice is left to the coastal State. The particular low-water line opted for is in turn dependent upon the choice of vertical level (technically the 'vertical datum') that represents the zero line from which heights and depths are measured. This reference level intersects the coast as the low-water line.

The higher the vertical datum is, the 'further up the coast' and thus more landwards the location of baselines will be. Similarly, the lower the vertical datum, the further 'down the beach' the low-water line is, and thus the starting line for the measurement of maritime claims is located further seawards. Unsurprisingly, States have tended to prefer the application of a low vertical datum to determine their low-water lines and thus normal baselines. In practice many coastal States favour the use of lowest astronomical tide (LAT) for this purpose. LAT is a particularly low vertical datum – thus low-water line – consisting of 'the lowest level which can be predicted to occur under average meteorological conditions and under a combination of astronomical conditions' (International Hydrographic Organization (IHO) et al. 2006, Chapter 2, p. 18). LAT is based on observations made over a period of 18.6 years 'identified as the relative rotation of the lunar and solar orbits or regression of the lunar nodes' (Sobey 2005, p. 65).

Unlike normal baselines that do not require declaration, other types of baselines such as straight, mouth of river closing lines and bay closing lines need to be explicitly declared. The baselines 'shall be shown on charts of a scale or scales adequate for ascertaining their position' or alternatively they can also be presented in 'a list of geographical coordinates of points, specifying the geodetic datum' (LOSC, Article 16 (1)). A geodetic datum (or 'horizontal datum', rather than a vertical datum as discussed above) is a reference from which measurements are made. Coastal States need to give due publicity to the charts or list of coordinates illustrating claimed baselines, providing a copy to the Secretary General of the United Nations (LOSC, Article 16 (2)). This also applies to the archipelagic baselines an archipelagic State can designate pursuant to the LOSC (LOSC, Article 47 (8–9)). With regard to the use of vertical datum, baselines other than normal baselines are also dependent upon the choice of this. For example turning points of straight and archipelagic baselines need to be anchored to the coast as represented by the low-water line. In short, the position of the low-water line will eventually affect the designation of baselines, irrespective of their types.

4. POTENTIAL IMPACTS ON MARITIME JURISDICTION

4.1 Ambulatory Baselines and Shifting Maritime Limits

It has long been recognized that parts of the coast are dynamic and can change location and configuration in relatively short periods. Indeed coastlines often change in a cyclical manner over time (alternately shifting seawards through deposition or accretion of material and then landwards because of erosion) (Hirst and Robertson 2004). Of particular note in the present context is that as the low-water line moves in response to sea-level rise, so the 'normal' baselines and the maritime claims measured from them will shift. This is fundamentally because normal baselines are coincident with the 'low-water line along the coast as marked on large scale charts officially recognized by the coastal State' (LOSC, Article 5). The location of normal baselines will therefore tend to move, or 'ambulate', over time in accordance with changes in the coast (Reed 2000; see also Prescott and Schofield 2005). The traditional, generally accepted implication of this phenomenon is that as the coast/normal baselines change, so will the maritime jurisdictional limits measured from them. Thus where the baseline advances (for example through the deposition of material along the coast) the outer limits of the maritime claims measured from that baseline will expand seawards. Conversely, where the normal baseline recedes (through coastal erosion) the coastal State may lose maritime areas as their maritime limits are pulled back.

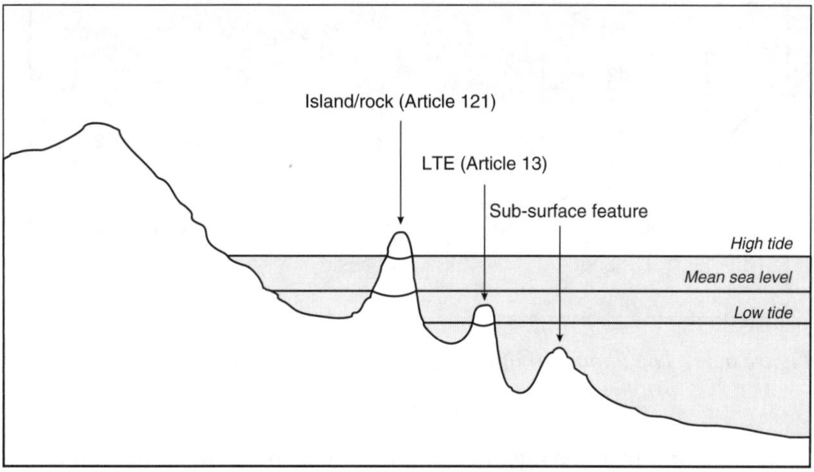

Figure 6.2 The impact of tidal levels on the status of insular features

Since normal baselines are represented by the low-water line, sea level is an important issue in their definition. This is a particularly significant issue as normal baselines are the predominant type of baseline worldwide. Moreover, while normal low-water line baselines would seem most obviously susceptible to change due to sea-level rise, as noted above other types of straight-line type baselines are also potentially threatened by sea-level rise because such baselines need anchoring to the coast, as represented by the low-water line.

Rising sea levels will predominantly lead to the retreat inland of the low-water line and thus the normal baseline. This can result in significant knock-on effects on the limits of maritime jurisdictional claims if the basepoints on which the limits of such claims depend similarly retreat inland. This threat to the extent of national maritime jurisdictional claims is especially significant for coastal States such as Bangladesh and parts of India in South Asia, as well as Vietnam in Southeast Asia, which have large stretches of low-lying coasts (Doyle et al. 2010). The maritime claims of States in possession (or even entirely composed) of low elevation islands such as Kiribati, the Marshall Islands and Tuvalu in the Pacific Ocean are also under threat from this phenomenon. Small, remote and low-lying islands can give rise to significant maritime jurisdictional entitlements. However sea-level rise could change the legal status of such insular features (see Figure 6.2). For example, an island that is currently always above the water surface even during high tide, may eventually disappear during high tide as a consequence of sea-level rise. This would lead to its

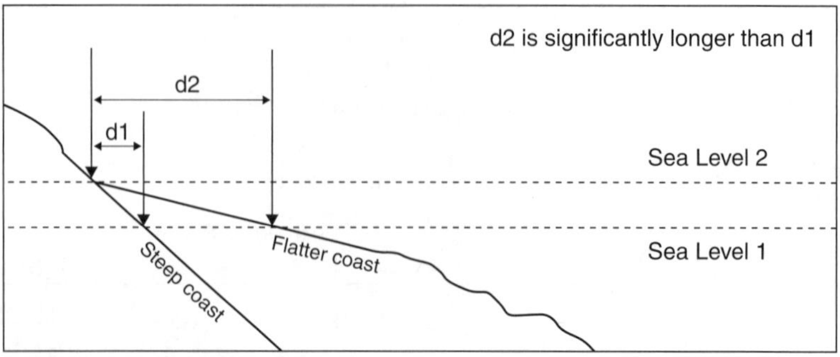

Figure 6.3 The impact of different sea levels on coasts of different gradients

reclassification from being an island, from which claims to the full range of maritime zones may be made under Article 121(2) of LOSC, to one of the categories of insular formations from which only restricted maritime claims can be made such as a 'rock' (LOSC, Article 121(3)) or a low tide elevation (LTE) (features that are exposed at low tide but are submerged at high tide) (LOSC, Article 13), or even a fully submerged feature that cannot be used to generate maritime claims (Schofield 2009).

4.2 Uneven Impacts

Although, as noted, sea-level rise would seem likely to result in the retreat of normal baselines inland, it is important to recognize the influence of coastal complexity and variability. Accordingly, sea-level rise is likely to result in uneven consequences in terms of impacts on maritime jurisdictional claims.

The gradient of the coast is a particularly important consideration in this context. Where the coastline is relatively steep the impact of sea-level rise will be limited in terms of shifting the location of baselines (and thus the maritime jurisdictional limits derived from them) horizontally. Conversely, where the coastline is gently shelving even relatively slight changes in sea level vertically can result in significant shifts in the location of the low-water line horizontally, and this in turn can have significant impacts on the spatial extent of national maritime claims.

For example, Figure 6.3 illustrates two different sea levels and two distinct coastline gradients. For the shallower gradient coastline shown, the impact of a rise in sea level from Level 1 to Level 2 is significantly more pronounced in terms of the horizontal change or recession inland in the location of the normal baseline. In short, the steeper the coastal area, the

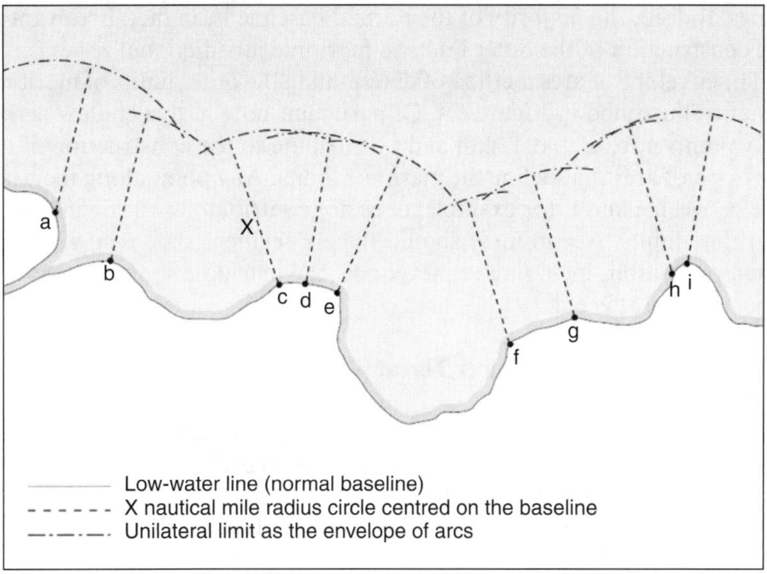

Figure 6.4 Basepoints, envelope of arcs and the limits of maritime jurisdiction

less horizontal impact there will be. A very shallow foreshore gradient –for example 5 per cent or less (around 3° measured from the horizontal) – will be affected by approximately 20 m of horizontal distance on the location of the normal baseline as a result of a 1 m difference in vertical datum (or sea level). For the same difference in vertical datum or sea level on a steep foreshore gradient – for example 173 per cent or more (around 60° measured from the horizontal) – can be affected only by 0.5 m of horizontal distance. In this context errors in the definition of the vertical datum can also affect the location of baselines (Leahy et al. 2001).

It is also worth emphasizing here that not all of a coastal State's baselines contribute towards the construction of the outer limits of its maritime claims. Maritime limits are commonly constructed through the 'envelope of arcs' method (Carleton and Schofield 2001). Consequently only certain basepoints along the normal baseline – essentially the outermost points along the baseline such as headlands and offshore islands – will be relevant to the limits of the maritime zones, with the length of the arcs ('X' on Figure 6.4) from the contributing basepoints determined by the breadth of the maritime zone for which the outer limit is being constructed. In contrast, those parts of the baseline that are located on the inner portion of a bay, for example, are unlikely to contribute to the outer limit of maritime zones. Indeed, the majority of the normal baseline is, in fact, irrelevant to the construction of the outer limits to maritime jurisdictional zones.

The envelope of arcs method of determining the outer limits of maritime zones is illustrated in Figure 6.4. Of particular note in this context is that only points a, b, c, d, e, f, g, h and i contribute to the construction of the envelope of arcs that define the maritime limits. Any point along the baseline between e and f, for example, does not contribute to the definition of maritime limits. Accordingly should the e–f segment shift landwards for whatever reason, including sea-level rise, the maritime limits in question would not be affected.

4.3 Islands and Coasts Under Threat

While large populations occupying low-lying coastal areas on continental coasts are arguably the most at risk from sea-level rise, it is noticeable that the debate on the issue tends to be framed, even dominated, by the concerns of and about a number of small island States. This focus, especially in the media narrative, may stem from a perception that in contrast to small low-lying island States, continental States have other, higher land to which displaced populations can retreat. Additionally the small island States are well placed to readily (and arguably justifiably) elicit sympathy for their apparent predicament, especially as they can argue convincingly that their contribution to global climate change through the emission of greenhouse gases has been negligible.

Concerns over sea-level rise by and on behalf of these States have been in large part prompted by recognition of not only certain States' limited land territory but also of how little of this is elevated above the present sea level. For example the highest point on the territory of the Maldives is only 2.4 m above sea level. The situation is similar in Tuvalu. For instance, His Excellency Mr Enele Sosene Sopoaga, former ambassador and permanent representative of the Mission of Tuvalu to the United Nations, suggested at the 2007 United Nations Framework Conference on Climate Change that sea-level rise is already a 'real emergency' for his country (Leake 2007). Consequently Tuvalu has been described as being at the 'front line of climate change' (Patel 2006). Analogous concerns also exist for other small relatively low-lying island States.

Moreover, it has been suggested that sea-level rise could ultimately lead to certain low-lying island States sinking beneath the waves, and thus losing their status as States. Under international law, codified in the Montevideo Convention on the Rights and Duties of States (signed at Montevideo, Uruguay, on December 26, 1933, hereafter the Montevideo Convention), States should possess a 'defined territory' and a 'permanent population', as well as a government and the capacity to enter into international relations with other States (Montevideo Convention, Article 1). The first two of these four requirements could be directly affected by sea-level rise. However, this scenario does not appear likely, at least in the near term. For instance, even if sea level was to rise by 1 m, even though the consequences of this would undoubtedly be calamitous, no State would be completely inundated as it stands. That said, even relatively slight sea-level rises might have major impacts on island habitability, for example impacting on the availability of potable water (Veitayaki et al. 2009). These concerns have led to the formation of bodies such as the Alliance of Small Islands States (AoSIS) which has been established to address issues of the vulnerability of small island States to climate change (AoSIS n.d.).

While there has been mounting evidence that the effects of sea-level rise are resoundingly negative, it has been speculated that sea-level rise might yield unlooked for benefits in terms of its impact on contentious territorial and maritime disputes. In particular, multiple territorial and maritime disputes in the Asia-Pacific, especially in the South China Sea and East China Sea, revolve around sovereignty over small, isolated and, critically, frequently low-lying islands. Such disputes, such as that concerning the Spratly Islands group in the South China Sea have proved to be long-standing source of friction among the multiple claimant states. The small insular features at the centre of these disputes may well be threatened with inundation through sea-level rise, potentially removing the key driver for these disputes, the territory at stake, from the equation.

The prospect of sea-level rise entirely submerging the fundamental focus of dispute, the islands themselves, and thus arguably resolving the conflict, represents an alluring prospect. Alternatively, sea-level rise might have the impact of reducing the disputed insular features to the status of mere rocks or low-tide elevations, thereby significantly curtailing their capacity to generate claims to maritime jurisdiction, and so serving to reduce or narrow the scope of the maritime jurisdictional dimensions of these disputes.

That said, States do not give up sovereignty claims readily. An example of this is the alleged disappearance of a disputed island called South Talpatty (by Bangladesh) or New Moore (by India), considered an inadvertent benefit of climate change. However this has not proven to be the case because not only were reports of the island's demise somewhat premature as it has been reported that the island still appears during 'very, very low tide conditions', but at least one of the parties to the dispute, Bangladesh, promptly reasserted its sovereignty claim to the feature despite its almost complete inundation (Wade 2009). In respect of other territorial disputes over low-lying islands it remains to be seen whether or not sea-level rise will yet have a positive impact on longstanding contentious island sovereignty disputes such as that over the Spratly Islands in the South China Sea or will add merely a further layer of confusion and doubt to an already uncertain scenarios (Dupont 2008).

Notwithstanding mounting evidence suggesting threats to islands and coasts due to climate change, counter-arguments do exist. For example there is evidence that coral atolls have proven to be remarkably robust over long periods, including periods when sea levels were considerably higher than they are now. This suggests that some insular features may be able to adapt naturally to climate change and sea-level rise. This is underpinned by observations in and analysis of 27 coral atoll islands in the central Pacific Ocean, which found that the majority (86 per cent) of these features had either remained stable or increased in area over a 20–60 year period despite reported sea-level rise in the central Pacific region (Webb and Kench 2010). Indeed it has been argued that uninterrupted sediment flows are necessary to sustain the island-building processes that maintain the integrity of coral reef islands: '[t]he physical dynamics of sediment supply and transport are critical factors in the context of management of rocky and sedimentary oceanic islands' (Kenchington 2009, p. 4). Accordingly, overpopulation of small islands, coupled with inappropriate land uses are important factors affecting the integrity of coral island ecosystems and thus the continued habitability of such features.

5. RESPONSE OPTIONS

Regardless of cause, it is evident that normal baselines are ambulatory, which, in turn, can shift maritime limits measured from them. However there is also a need to have jurisdictional clarity for better ocean space management and thus fixed maritime limits. Three main response options arise in this context: relocation, protecting the coast or stabilizing baselines, and/or fixing the position of outer maritime limits.

5.1 Planned Retreat and Relocation

Rather than attempting to protect the coast and stabilize its present location (see below), one alternative would be to, in a sense, accept the inevitable and manage the impacts of rising sea level. This can be achieved through approaches that recognize increased dynamism in the coastal zone and thus provide for coastal development that is responsive rather than resistant to change. In this context 'planned retreat' calls for coastal development to be removed/relocated once defined 'setbacks' are encroached on through coastal erosion (Smith et al. 2011).

A more extreme scenario envisages the abandonment of entire islands. For example the Indian island of Lohachara located in the Sundarbans region where the Ganges and Brahmaputra rivers empty into the Bay of Bengal and once home to 10 000 people, was reportedly evacuated due to the effects of sea-level rise, although this was done 'as a precaution' (Lean 2006). Similarly, a decision was made in 2005 to relocate the 2600 inhabitants of the Carteret Islands of Papua New Guinea (International Organization for Migration 2009). It can be anticipated that such responses will increasingly come to the fore if predictions regarding substantial and rapid sea-level rise prove to be well founded.

5.2 Stabilizing the Baselines

To stabilize baselines, physical and legal approaches can be attempted. There has been a long history of human efforts in protecting valuable parts of the coastline and thus, often incidentally, in stabilizing portions of the baseline along the coast for maritime jurisdictional purposes. Such efforts tend to involve the building of sea defences such as sea walls, groynes and wave reduction structures. Such efforts are intended to prevent or at least delay natural processes of erosion and abrasion (Freestone 1991; Schofield 2009).

Similarly, reclamation could also be an option for building up vulnerable coastlines. For example, the Republic of Maldives has started projects to build up some of its large islands through reclamation to ensure that it will have more safe refuges for its population (Morris 2009). For small islands physical intervention can serve as a means to protect insular status. The classic case is that of Japan's southernmost territory, the small, isolated insular feature, Okinotorishima, around which Japan has constructed a 360° sea wall, vertically higher than the threatened feature extends above sea level (Prescott and Schofield 2005).

The drawbacks of the physical approach described are that it may be environmentally unfriendly and is also frequently costly. For instance, the above-mentioned structures built for Okinotorishima cost in excess of US$200 million in the 1980s (Brown et al. 1991). This is certainly not a preferred option for less developed States like the Maldives, Tuvalu or other Small Island Developing States (SIDS), which appear likely to suffer most from the impact of sea-level rise. This is particularly the case in the context of long, narrow coral atolls (Freestone 1991).

Physical intervention may also disturb the natural equilibrium of a coast and interrupt sediment flows, leading to serious 'unintended consequences' for the environment in the long term (Kenchington 2009). The building of coastal defence structures can also affect ecologies on local and regional scales, for example by affecting the existence of species and thus changing the native assemblages of surrounding areas (Airoldi et al. 2005). It can therefore be concluded that the physical intervention approach to stabilizing baselines is generally costly and tends to be environmentally and ecologically unfriendly. In addition, physical interventions such as reclamation may spark legal questions concerning the validity of reclaimed coastlines to be used as baselines. An ecosystem-based and sustainable management approach is therefore strongly advocated (Kenchington 2009).

The other option for 'fixing' baselines is to take what can be characterized as a legal approach. This has been suggested as a means whereby States threatened by sea-level rise might be able to retain their maritime claims (Soons 1990). Two major alternatives to this are fixing the normal baselines or fixing the maritime limits. LOSC states that the normal baselines of coastal States are the low-water lines depicted on a nautical chart recognized by the coastal States. The key information in this LOSC article is that 'a recognized nautical chart' is the legal document on which the normal baselines of a coastal State are declared. However, there is no clause in the article detailing the required technical specifications of the nautical chart. For example, the article does not specify either the age of the chart in question or whether it needs to be registered or recognized by an international body.

On the aforementioned basis it has been observed that the coastal State may use any chart in defining its normal baselines as long as the chart is officially recognized by the coastal State itself (Schofield 2009). Arguably a coastal State could therefore fix its normal baselines by recognizing a chart showing such baselines. However, if the baselines were to subsequently move, the coastal State would need to produce different charts officially recognized for different purposes – that is, charts for illustrating baselines as well as those used for navigational purposes. For the latter, a nautical chart has to be regularly revised through surveys to show the most updated coastal environment and important objects, especially those hazardous to navigation.

The potential issue with the use of specific and fixed charts showing baselines is that other States may not necessarily recognize the chart chosen by the coastal State. This can be problematic if two States need to delimit maritime boundaries between them. If one State fails to recognize another State's chart depicting normal baselines, the progress of the delimitation may be hampered. However, it is not unusual for two States to agree on the use of a particular chart for maritime delimitation, even though the chart may no longer depict the current coastline/baselines. A good example of this practice is the maritime boundary delimitation between Indonesia and Singapore in the Singapore Strait, signed on 10 March 2009 (Republic of Indonesia, Ministry of Foreign Affairs 2009). Indonesia and Singapore agreed that the delimitation line be constructed using charting that does not show the major recent changes to Singapore's coastline as a result of reclamation works (Republic of Indonesia 2010). To anticipate problems caused by disagreement on the use of fixed baselines depicted by a particular chart, coastal States can voluntarily declare their fixed normal baselines in the same manner as States usually declare straight or archipelagic baselines. By doing this, protest and disagreement from other States, usually neighbours, can be anticipated well in advance, before the baselines are used for maritime claims and, particularly, delimitation.

The potential instability of normal baselines may also be partially overcome by the use of straight baselines. However, as noted above, straight baselines were not originally conceived with the aim of fixing baselines in response to global changes such as sea-level rise. Straight baselines can be used in the context of deltas and unstable coasts (LOSC, Article 7 (2)). It is questionable whether they could be applied along an entire coastline as a response to sea-level rise, raising questions about the validity of such straight baselines. Further, as noted, straight baselines still require turning points, which should be points somewhere at the interface of land and water during low tide. Straight baselines cannot, therefore, 'float' out to sea, unattached to any point of land, but must be bound back to terrestrial anchor points. Accordingly straight baselines still require the use of base-points, the location of which depends on the choice of low-water line, which is characteristically unstable. This implies that the use of straight baselines may fix baselines in a particular location or situation, but not fully resolve issues of baseline instability. Another issue with straight baselines is that while many States tend to interpret Article 7 of the LOSC liberally in designating straight baselines, since there are some uncertainties and ambiguities therein, other States, including major maritime powers such as the United States, have resisted this trend. Thus, straight baselines claimed in this manner are liable to being considered excessive by other States and may therefore be opposed and protested as a result.

5.3 Fixing the Maritime Limits

Fixing maritime limits may be an alternative to stabilizing baselines for dealing with changing baselines due to climate change. This would mean that once maritime limits are set they are permanent in terms of location. Hence it would not matter whether coastlines or baselines shift due to sea-level rise, maritime limits would stay where they were. Should this be adopted States will not be disadvantaged if there is significant sea-level rise that shifts baselines landwards. However some States may not see this as an advantageous option if for some reason their baselines shift further seawards. This is possible because of deposition of material along the coast for example. Ironically, as noted, this appears to have occurred with reference to some Pacific islands, 'largely due to coral debris, land reclamation and sediment' (Webb and Kench 2010). If coastlines or baselines shift further seawards but maritime limits remain fixed, this would be a less than ideal scenario for the States involved. However considering the prevailing perception that sea-level rise is accelerating, threats to the location of baselines and thus the scope of maritime claims do still exist. This is especially the case since it is unclear whether or not the natural responses that, for example, coral islands have exhibited in the past in response to sea-level variability will be able to cope with the potentially rapid sea-level rise induced by global climate change. This may make the option of fixing maritime limits more attractive to island States in the Pacific vulnerable to climate change.

The key ways in which the limits of maritime jurisdiction are determined are explored in the next section.

6. SETTING THE LIMITS OF MARITIME JURISDICTION

The limits of a coastal State's maritime jurisdiction can be established in one of three ways. First, maritime claims can be generated to the full extent or distance allowed under international law in the absence of analogous claims on the part of neighbouring States. Second, where overlapping claims to maritime jurisdiction exist maritime boundaries may be delimited between neighbouring States. Third, the definition of the outer continental shelf limits involves a submission process to the United Nations Commission on the Limits of the Continental Shelf (CLCS) (see below).

Provided a coastal State does not have any neighbours with an overlapping claim to a particular zone, it can define its maritime limits unilaterally. For example, if a coastal State has no neighbours within 24 nm of its baselines it can unilaterally define its territorial sea limits. With regard to this option, the outer limits of maritime zones are commonly defined using the method of envelope of arcs outlined previously. The limits of such claims are generally dependent on the baselines from which these maritime claims are measured, thus they may move over time as baselines shift. However, it is worth noting that the use of this method employs only relevant basepoints along baselines to generate maritime limits. Depending on the shape or configuration of a coastline and therefore its associated baselines, not every point along baselines will affect the location of maritime limits. In other words while one part of the baselines may be crucial in constructing maritime limits, other parts may be irrelevant. However, it is generally true that baseline changes can shift maritime limits.

The introduction of 200 nm breadth EEZs in particular has had a dramatic effect on the scope of ocean spaces becoming subject to the maritime claims of coastal States. It has been estimated that if every coastal State made national maritime jurisdictional claims out to 200 nm (as is predominantly the case) these claims would encompass 43 million nm^2 (147 million km^2) of maritime space. This amounts to approximately 41 per cent of the area of the oceans or 29 per cent of the earth's surface. The extent of the area subject to jurisdictional claims out to 200 nm is thus approximately equivalent to the area of land territory on the surface of the earth (Prescott and Schofield 2005; Symonds 2011). The inevitable consequence of this enormous expansion in national claims to maritime space seawards has been a major proliferation in overlapping claims to maritime jurisdiction and thus potential international maritime boundaries. Indeed there is presently no coastal State in existence that can claim a full suite of maritime zones without overlapping claims. For example, to claim a full 200 nm EEZ, a coastal State must be over 400 nm from its nearest maritime neighbour. Thus, every coastal State needs to delimit at least one maritime boundary. However, this situation does not affect the way the breadth of maritime zones are measured from baselines.

Considering the geographical location of coastal States in the world and the configuration of their coasts, overlapping claims of maritime zones among coastal States is inevitable (see Figure 6.5). As such, maritime delimitation is required to produce maritime boundaries. Maritime delimitation among States is therefore another way for coastal States to define the limits of their maritime zones. While the first option is a unilateral process, maritime delimitation is a bilateral or multilateral one.

The delimitation of maritime boundaries between two or more coastal States is governed by the principles and rules of public international law (Prescott and Schofield 2005). International law explains how maritime boundary delimitations should be established. However maritime boundary delimitation is usually resolved either through negotiation among the affected parties or by submission of the case to a third party. This third party can be arbitrators, mediators, courts or tribunal such as the International Court of Justice (ICJ) or the International Tribunal for the Law of the Sea (ITLOS).

As Figure 6.5 illustrates, an overlapping EEZ and continental shelf claim exists where the distance between State A and State B is less than

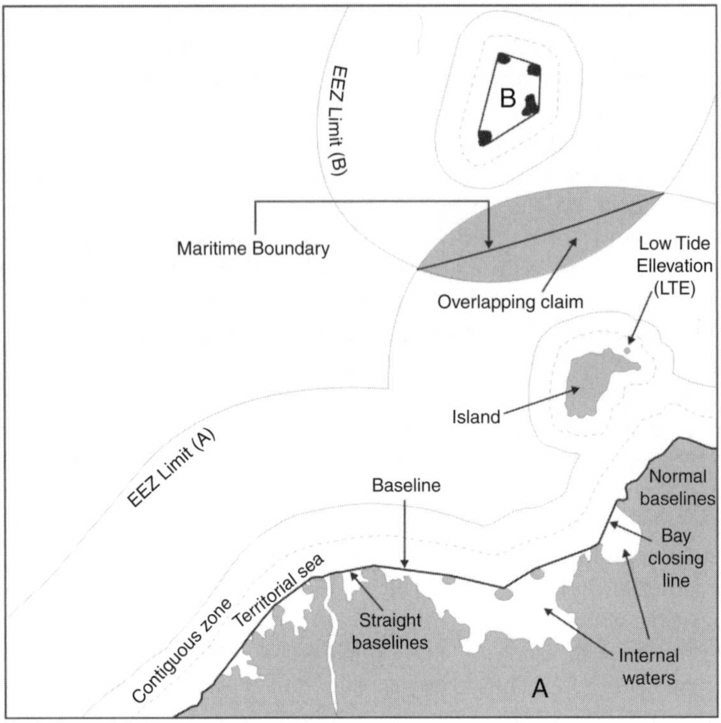

Figure 6.5 Maritime limits and boundaries

400 nm but greater than 24 nm. If the distance between the two neighbouring States is less than 24 nm then their territorial seas will overlap. This illustrates that maritime boundary delimitation can be required for territorial sea, EEZ or continental shelf depending on the distance between the States in question. In this situation the rules governing maritime boundary delimitation for those different zones are also different. For the territorial sea, for example, it is explicitly stated by the LOSC that 'neither of the two opposite or adjacent States is entitled to extend its territorial sea beyond the median line' unless either State involved agrees otherwise or if 'historic title or other special circumstances' exist (LOSC, Article 15). A median line or equidistance line has been defined as 'a line every point of which is equidistant from the nearest points on the territorial sea baselines of two States' (IHO et al. 2006, Chapter 6, p. 3). However, the LOSC does not specifically mention methods for delimiting EEZ and continental shelf boundaries in case overlapping claims between two or more States are identified. Instead the relevant provisions of the LOSC only mention

that continental shelf and EEZ boundaries between States with opposite or adjacent coast should be established to 'achieve an equitable solution' (LOSC, Articles 74 and 83).

One important issue in this context is that once international boundaries are established they tend to stay where they are. Maritime boundaries do not change unless the parties in question agree. The 1969 Vienna Convention on the Law of Treaties states that boundary treaties are excluded from the rule that a party to a treaty may invoke 'a fundamental change in circumstances' as grounds for terminating a treaty (United Nations 1969, Article 62 (2a)). In addition, the 1978 Vienna Convention on Succession of States in Respect of Treaties also provides that a change of States does not affect a boundary established by a treaty (United Nations 1978, Article 11 (a)). In other words, agreed maritime boundaries are fixed in terms of location, even if the baselines they are constructed from have shifted.

The third option for defining the outer limits of maritime jurisdiction is through submission to a third party, as in the case of continental shelf beyond 200 nm from baselines. To confirm its sovereign rights over areas of continental shelf beyond 200 nm from its baseline, the LOSC provides that a coastal State should make a submission regarding its proposed outer continental shelf limits mainly based on geological and geomorphologic evidence and submit this to the CLCS. The 'continental shelf beyond 200 nm from baseline' is commonly termed the 'outer' or 'extended' continental shelf (see Figure 6.1). Determining the outer limit of the continental shelf where it extends seawards of the 200 nm from baselines involves complicated procedures and significant resources. The procedure for the delineation of the outer limits of continental shelf beyond 200 nm was further specified by the CLCS in its *Scientific and Technical Guidelines*, which were adopted on 13 May 1999 (CLCS 1999). Once the CLCS has delivered its recommendations to the coastal State that State may declare the outer limits of its outer continental shelf, which are 'final and binding' when defined 'on the basis of' the CLCS's recommendations (LOSC, Article 76 (8)). Consequently, even though the outer limit of the continental shelf is not definitive in terms of distance from baselines, unlike the outer limits of other zones the limit is fixed in term or location once it has been properly established.

From the three options to set the limits of maritime jurisdiction elaborated here, the first generates unfixed maritime limits while the latter two establish fixed limits. The shifting maritime limits in the first option result from migrating baselines, especially normal baselines. The inherent dynamism of the coast will inevitably lead to alterations in the location of baselines over time, and this in the present circumstances will necessarily lead to changes in the location of the outer limits of maritime claims. This situation is likely to be exacerbated by sea-level rise. In the second and third options, agreed maritime boundaries and outer limits of continental shelf beyond 200 nm are fixed. The advent of sea-level rise has led to a growing desire on the part of threatened coastal States to fix baselines and the limits derived from them. In this context it can be observed that the challenge of global sea-level rise was simply not contemplated during the drafting of LOSC. The fact that in accordance with LOSC some maritime limits and boundaries should be fixed does, however, suggest that moving towards declaring and fixing maritime limits presently susceptible to change represents a plausible and reasonable response to an unanticipated problem.

7. CONCLUSIONS

Sea-level rise has significant potential to have highly problematic effects not just in terms of the inundation of land territory but also with respect to the extent of coastal State claims to maritime space. This chapter has explored some of the issues and uncertainties that arise in relation to the phenomenon of sea-level rise. It has also sought to highlight some of the ways the impacts of sea-level rise on claims to maritime jurisdiction are likely to be uneven.

A number of options to address this challenge have been outlined. While physical intervention with a view to stabilizing or fixing coastlines/baselines and thus maritime limits may work in certain circumstances and situations, the legal fixing of either or both baselines and limits may prove a more feasible solution. This can begin from coastal States unilaterally declaring or even depositing their normal baselines or maritime limits, analogous to the deposition of straight (LOSC, Article 16 (20)) or archipelagic (LOSC, Article 47 (9)) baselines. Once declared on an official chart, normal baselines can remain at the same location until the chart is revised. In addition, normal baselines are those identified 'on large-scale charts officially recognized by the coastal State' (LOSC, Article 5) and are not necessarily representative of the actual location of coastlines. This supports the idea of fixing normal baselines by preserving charts assigned for the purpose of depicting baselines. Increasing State practice along these lines can be anticipated.

A more radical approach to fixing baselines would be to amend the LOSC. However, this approach seems unlikely because even though the LOSC contains amendment procedures they have never been activated. Further, there appears to be scant enthusiasm for a Fourth United Nations Conference on the Law of the Sea. Such discussion could nevertheless potentially take place through consultations in a technical forum, at least initially, such as the Advisory Board on the Law of the Sea (ABLOS) (Schofield 2009). In line with this idea, a supplementary agreement to LOSC, such as the fish stocks agreement of 1995 (United Nations 1995), is conceivable. This precedent could be used as a model in approaching the fixing of baselines and/or maritime limits.

Fixing baselines and/or maritime limits is essentially aimed at preserving the existing rights of coastal States. Therefore it can be argued that this is hardly excessive. In addition this effort is particularly important for small island States which have minimal responsibility for the emergence of the problems related to climate change. The Small Island Developing States (SIDS) have contributed least to the human-induced climate change that is causing sea-level rise but are affected most by its impacts. It seems only equitable that such States should be able to fix their baselines and maritime limits to preserve their rights over their maritime zones and natural resources to which they are entitled.

REFERENCES

Airoldi, L., M. Abbiati, M.W. Beck, S.J. Hawkins, S.J. Jonsson, D. Martin, P. S. Moschella, A. Sundelöf, R.C. Thompson and P. Åberg (2005), 'An ecological perspective on the deployment and design of low-crested and other hard coastal defence structures', *Coastal Engineering*, **52** (10–11), 1073–1087.

Alam, Mahmood (1996), 'Subsidence of the Ganges-Brahmaputra Delta of Bangladesh and associated drainage, sedimentation and salinity problems', in John D. Milliman and Bilal U. Haq (eds), *Sea-Level Rise and Coastal Subsidence: Causes, Consequences, and Strategies*, Dordrecht, The Netherlands: Kluwer Academic Publishers, pp. 169–192.

AoSIS (n.d.), Website, accessed 21 April 2012 at: http://www.sidsnet.org/.

Ananthaswamy, A. (2009), 'Going, going . . . ', *New Scientist*, 4 July, pp. 26–31.

Australian Government, Department of Climate Change (2009), *Climate Change Risks to Australia's Coast: a First Pass National Assessment*, 14 November, last accessed 21 October 2010 at: http://www.climatechange.gov.au/~/media/publications/coastline/cc-risks-full-report.pdf.

Brown, J., A. Colling, D. Park, J. Phillips, D. Rotehery and J. Wright (1991), *Case Studies in Oceanography and Marine Affairs*, Open University Oceanography vol. 6, Milton Keynes, UK: Pergamon Press in association with the Open University.

Carleton, Chris and Clive Schofield (2001), *Developments in the Technical Determination of Maritime Space: Charts, Datums, Baselines, Maritime Zones and Limits*, Maritime Briefing vol. 3, no. 3, Durham, UK: International Boundaries Research Unit.

Church, John A., John R. Hunter, Kathleen L. McInnes and Neil J. White (2006), 'Sea-level rise around the Australian coastline and the changing frequency of extreme sea-level events', *Australian Meteorological Magazine*, **55**, 253–260.

Church, John A., Neil J. White, Thorkild Aarup, W. Stanley Wilson, Philip L. Woodworth, Catia M. Domingues, John R. Hunter and Kurt Lambrek (2008), 'Understanding global sea levels: past, present and future', *Sustainability Science*, **3** (1), 9–22.

CLCS (1999), *Scientific and Technical Guidelines of the Commission on the Limits of the Continental Shelf*, CLCS/11, 13 May, Geneva, Switzerland: United Nations, accessed 21 April 2012 at: http://www.un.org/depts/los/clcs_new/documents/Guidelines/CLCS_11.htm.

Douglas, Bruce C. (1997), 'Global sea rise: a redetermination', *Surveys in Geophysics*, **18** (2–3), 279–292.

Doyle, Thomas W., Richard H. Day and Thomas C. Michot (2010), *Development of Sea Level Rise Scenarios for Climate Change Assessments of the Mekong Delta, Vietnam*, Reston, Virginia: U.S. Geological Survey Open-File Report 2010-1165.

Dupont, Alan (2008), 'The strategic implications of climate change', *Survival: Global Politics and Strategy*, **50** (3), 29–54.

Freestone, David (1991), 'International Law and Sea Level Rise', in Robin R. Churchill and David Freestone (eds) *International Law and Global Climate Change*, London/Dordrecht: Graham and Trotman/Martinus Nijhoff, pp. 119–122.

Gornitz, Vivien (1995), 'Monitoring sea level changes', *Climatic Change*, **31** (2–4), 515–544.

Hansen, Jim (2006), 'The threat to the planet', *The New York Review of Books*, 13 July, accessed 21 April 2012 at: http://www.nybooks.com/articles/19131.

Hirst, Bill and David Robertson (2004), 'GIS, charts and UNCLOS – can they live together?', *Maritime Studies*, **136** (May–June), 1–6.

IHO with the International Oceanographic Commission and the International Association of Geodesy (2006), *A Manual on Technical Aspects of the United Nations Convention on the Law of the Sea – 1982*, Special Publication No. 51, 4th edition, Monaco: International Hydrographic Bureau.

IPCC (2007), 'Causes of change', *IPCC Fourth Assessment Report: Climate Change 2007*, Geneva, Switzerland: IPCC, accessed 21 April 2012 at: http://www.ipcc.ch/publications_and_data/ar4/syr/en/spms2.html.

International Organization for Migration (IOM) (2009), *IOM Policy Brief: Migration, Climate Change and the Environment*, Geneva, Switzerland: IOM, accessed 21 April 2012 at: http://www.preventionweb.net/files/9845_policybrief.pdf.

Kenchington, Richard (2009), 'Maintaining coastal and lagoonal ecosystem and productivity', in Hiroshi Terashima (ed.), *Proceedings of International Symposium on Islands and Oceans: January 22 and 23, 2009*, Tokyo, Japan: Ocean Policy Research Foundation, pp. 1–11, accessed 21 April 2012 at: http://www.sof.or.jp/en/report/pdf/200903_ISBN978-4-88404-217-2.pdf.

Leahy, Frank J., Brian A. Murphy, Philip A. Collier and David J. Mitchell (2001), 'Uncertainty issues in the geodetic delimitation of maritime boundaries', in *Proceedings of the 2001 ABLOS Conference, Monaco, 18–19 October 2001*, CD-ROM, accessed 21 April 2012 at: http://www.gmat.unsw.edu.au/ablos/ABLOS01Folder/LEAHY.PDF.

Leake, J. (2007), 'Global warming and the world's low-lying countries', *The Sunday Times*, accessed 21 April 2012 at: http://www.timesonline.co.uk/tol/news/environment/article3054040.ece.

Lean, Geoffrey (2006), 'Disappearing world: global warming claims tropical island', *The Independent*, 24 December, accessed 21 April 2012 at: http://www.independent.co.uk/environment/climate-change/disappearing-world-global-warming-claims-tropical-island-429764.html.
Mercer, John H. (1968), 'Antarctic ice and Sangamon sea level', *International Association for the Hydrological Sciences Symposium*, **79**, 217–225.
Mercer, John H. (1978), 'West Antarctic ice sheet and CO_2 greenhouse effect: a threat of disaster', *Nature*, **271**, 321–325.
Montevideo Convention on the Rights and Duties of States, opened for signature 26 December 1933, 165 LNTS 19 (entered into force 26 December 1934), Montevideo, Uruguay.
Morris, Chris (2009), 'Maldives rises to climate challenge', *BBC News*, 17 March, accessed 21 April 2012 at: http://news.bbc.co.uk/2/hi/7946072.stm.
Oppenheimer, Michael and R.B. Alley (2004), 'The West Antarctic ice sheet and long term climate policy', *Climatic Change*, **64** (1), 1–10.
Patel, Samir S. (2006), 'A sinking feeling', *Nature*, **440**, 734–736.
Prescott, J.R. Victor and Clive H. Schofield (2005), *The Maritime Political Boundaries of the World*, Leiden, The Netherlands and Boston, MA, USA: Martinus Nijhoff Publishers.
Reed, M. (2000), *Shore and Sea Boundaries: The Development of International Maritime Boundary Principles through United States Practice*, Washington DC, USA: US Department of Commerce.
Republic of Indonesia (2010), 'Working meeting of Commission One of DPR on the ratification of the agreement between the Republic of Indonesia and the Republic of Singapore regarding maritime boundary delimitation between the two states in respect to the western area of the Singapore Strait' (in Indonesian), press release, 24 May, Jakarta: Government of the Republic of Indonesia, accessed 21 April 2012 at http://www.deplu.go.id/Pages/SpeechTranscriptionDisplay.aspx?Name1=Pidato&Name2=Menteri&IDP=646&l=id.
Republic of Indonesia, Ministry of Foreign Affairs (2009), 'The signing of the treaty between the Republic of Indonesia and the Republic of Singapore relating to the delimitation of the territorial seas in the western part of the Strait of Singapore', press release, 10 March, Jakarta, Indonesia: Ministry of Foreign Affairs, accessed 21 April 2012 at: http://www.kemlu.go.id/Pages/PressRelease.aspx?IDP=646&l=en.
Schofield, Clive H. (2009), 'Shifting limits? Sea level rise and options to secure maritime jurisdictional claims', *Carbon and Climate Law Review*, **3** (4), 405–416.
Schofield, Clive H. (2011) 'Rising waters, shrinking states: the potential impacts of sea level rise on claims to maritime jurisdiction', *German Yearbook of International Law*, 53 (2010), 189–231.
Smith, T.F., S. Myers, D.C. Thomsen and J. Rosier (2011), 'Integrated coastal zone management and planning', in Warwick Gullett, Clive H. Schofield and Joanna Vince (eds), *Marine Resource Management*, Chatswood, Australia: LexisNexis Butterworths, pp. 109–121.
Sobey, Rodney J. (2005), 'Extreme low and high water levels', *Coastal Engineering*, **52** (1), 63–77.
Soons, A.H.A. (1990), 'The effects of sea level rise on maritime limits and boundaries', *Netherlands International Law Review*, **37** (2), 207–232.
Sutanta, Heri, Abbas Rajabifard and Ian D. Bishop (2009), 'An integrated

approach for disaster risk reduction using spatial planning and SDI platform', in B. Ostendorf, P. Baldock, D. Bruce, M. Burdett and P. Corcoran (eds), *Proceedings of the Surveying and Spatial Sciences Institute Biennial International Conference*, Adelaide, Australia: Surveying and Spatial Sciences Institute, p. 341.

Symonds, P.A., Senior Adviser at Law of the Sea, Geoscience Australia (2011), personal communication, July.

United Nations (1969), Vienna Convention on the Law of Treaties, opened for signature 23 May 1969 (entered into force 27 January 1980), UN Doc. A/Conf.39/27; 1155 UNTS 331; 8 ILM 679 (1969); 63 AJIL 875 (1969), Geneva, Switzerland: United Nations, accessed 22 April 2012 at: http://untreaty.un.org/ilc/texts/instruments/english/conventions/1_1_1969.pdf.

United Nations (1978), Vienna Convention on Succession of States in Respect of Treaties, opened for signature 23 August 1978 (entered into force 6 November 1996), 1946 UNTS 3, Geneva, Switzerland: United Nations, accessed 22 April 2012 at: http://untreaty.un.org/ilc/texts/instruments/english/conventions/3_2_1978.pdf.

United Nations (1983), United Nations Convention on the Law of the Sea, opened for signature 10 December 1982, Montego Bay, Jamaica (entered into force 16 November 1994), publication no. E97.V10, New York, NY, USA: United Nations, see 1833 UNTS 3.

United Nations (1995), Agreement for the Implementation of the Provisions of the United Nations Convention on the Law of the Sea of 10 December 1982 Relating to the Conservation and Management of Straddling Fish Stocks and Highly Migratory Fish Stocks (adopted 4 August 1995, in force as from 11 December 2001), Geneva, Switzerland: United Nations, accessed 22 April 2012 at: http://daccess-dds-ny.un.org/doc/UNDOC/GEN/N95/274/67/PDF/N9527467.pdf?OpenElement.

United Nations Development Program (UNDP) (2011), 'Vietnam: Sea level rise could "displace millions"', 20 May 2011, accessed 21 April 2012 at: http://www.undp.org.vn/digitalAssets/25/25093_VIETNAM_sea_level_risex.pdf.

Vaughan, David G. (2008), 'West Antarctic ice sheet collapse – the fall and rise of a paradigm', *Climatic Change*, **91** (1–2), 65–79.

Veitayaki, J., P. Manoa and A. Resture (2009) 'Pacific Islands and the problems of sea level rise due to climate change', in Hiroshi Terashima (ed.), *Proceedings of The International Symposium of Islands and Oceans*, Tokyo: Ocean Policy Research Foundation, pp. 55–69.

Wade, Matt (2009) 'Rising sea level settles border dispute', *The Sydney Morning Herald*, 25 March, accessed 22 April 2012 at: http://www.smh.com.au/environment/climate-change/rising-sea-level-settles-border-dispute-20100324-qwum.html.

Walker, Gabrielle and David King (2008), *The Hot Topic: How to Tackle Global Warming and Still Keep the Lights On*, London, UK: Bloomsbury Publishing.

Webb, Arthur P. and Paul S. Kench (2010), 'The dynamic response of reef islands to sea-level rise: evidence from multi-decadal analysis of island change in the Central Pacific', *Global and Planetary Change*, **72** (3), 234–246.

Woodroffe, C.D. (2011) 'The Australian margin: ocean and coastal processes', in Warwick Gullett, Clive H. Schofield and Joanna Vince (eds), *Marine Resource Management*, Chatswood, Australia: LexisNexis Butterworths, pp. 9–21.

7. Climate change and maritime security

Stuart Kaye

1. INTRODUCTION

The first decade of the twenty-first century has seen a dramatic increase in international interest in two key issues facing the international community. The first of these concerns, which grew to prominence in the wake of the terrorist attacks in the United States on 11 September 2001, is the international focus on maritime security. The introduction of the International Ship and Port Facility Security Code (ISPS Code 2004) within the International Maritime Organization (IMO) in 2002, the negotiation and continuing implementation of the Proliferation Security Initiative (PSI) from 2003 (Kaye 2005, pp. 205–229), the adoption of unilateral measures such as the Container Security Initiative by the United States, and most recently the renegotiation of the Convention for the Suppression of Unlawful Acts Against the Safety of Maritime Navigation (SUA Convention 1988; SUA Protocol 2005) in October 2005, have all occurred in direct response to the threat of terrorism to shipping. In the foreseeable future, further likely amendments to the SUA Convention, the continuing evolution of rules on Long Range Identification and Tracking of vessels using satellite monitoring technology within the IMO, and increasing regional cooperation through the Asia Pacific on maritime security are all possible.

The second issue, the international community's responses to climate change has also been extremely prominent, albeit over a longer period. The international legal outcomes have included the United Nations Framework Convention on Climate Change (UNFCC 1992) and its Kyoto Protocol (Kyoto Protocol 1997), and the Ozone Convention (Ozone Convention 1985), as well as more recently the major international conference at Copenhagen (COP-15 2009).The predictions of the Intergovernmental Panel on Climate Change (IPCC) of the United Nations in its periodic reports include a future where there is increased desertification for some areas, inundation of coastal regions in others, as well as increasing variability in climate disrupting agricultural production (IPCC 4th Report 2007, esp. chap. 6 and 15).

While these issues, security and climate change are clearly important to the international community in a contemporary context, it is not immediately obvious that the two concepts are related or that they have much in the way of overlap. However, this chapter will consider how maritime security and climatic change are related, and why the consequences of the latter may increasingly inform the nature of responses to the former. The areas of potential overlap are distinct, and each will be dealt with in turn.

2. RESOURCE PROTECTION

As considered in an earlier chapter, climate change will have an impact upon the marine environment and therefore on marine living resources. Access to such resources is in the national interest and is thus linked to security. While climate change is most often discussed in the media in terms of variations in the content of gases in the atmosphere, resulting in changes to weather patterns over time, it would be wrong to discount the possibility of change in the world's ocean currents and to water, salinity and acidity temperature.

Alterations in long-term rises in air temperature could, for example, see the melting of large areas of ice, such as the Greenland Ice Sheet, altering water salinity levels in high latitudes, which could in turn affect ocean currents and marine living resource habitats. Terrestrial and marine pollution may cause a rise in ocean acidity levels, which may be exacerbated by increased take up of carbon dioxide from the atmosphere by the world's oceans (Feely et al., 2004, pp. 362–366). Changes in water acidity, temperature and salinity, as well as ocean currents are likely to have a significant impact upon fisheries. Some species are already known to move within a habitat range depending on a variety of factors for reasons that are not always well understood (Winberg et al 2012). A realistic outcome of climate change in the longer term may therefore be the displacement and movement of many commercially valuable marine living resource stocks (Rahman 2012).

The implications of a movement of fish stocks are significant from an ecological point of view. Fisheries represent a significant part of the diet of billions of people, and the movement or decline of key stocks could have catastrophic impacts upon the well-being of hundreds of millions of people, most notably in East and South Asia, coastal Africa and South America. The Food and Agriculture Organization (FAO) has calculated that world wild fishery output has essentially plateaued over the past decade at between 80 and 85 million tonnes per annum (FAO 2007). While this figure has been supplemented by freshwater fishing and aquaculture, it still represents the majority of fish production. Even the rises in aquaculture production, estimated to be around 10 million tonnes between 2000 and 2004, may not be sufficient to offset any substantial decline in capture fisheries. Nor can it be assumed that aquaculture would be immune from the environmental pressures such as salinity, acidity and water temperature that could adversely affect wild fisheries.

The pressure on fisheries could also be increased as a result of climate change. Desertification of some areas could see significant declines in terrestrial food production. If a large agricultural region receives less rainfall, then it is reasonable to expect long-term food production from such a region to decrease. The 4th IPCC Report estimated world food production would suffer an overall decrease with a temperature increase of in excess of 3 degrees Centigrade (IPCC 4th Report 2007). This in turn may see efforts to replace the lost protein from agricultural production shifted to fisheries, which will place world fisheries under greater strain.

While a definition of maritime security that encompasses food security may seem to some like a long bow to draw, it is not unreasonable to make the argument. If world terrestrial food production drops, greater pressure on fisheries would lead to a host of problems more familiar to those with an interest in maritime security. Greater pressure on specific fisheries in parts of the world in the present has typically led to illegal fishing in the maritime zones of neighbouring States, and there is no reason to assume that such illegal fishing would not occur in the future. Such developments would inevitably lead to maritime security challenges and, in all likelihood, bilateral disputes arising.

An excellent example of this was the significant increase in illegal fishing that took place in the waters of northern Australia in 2006–2007. It was reported that fish stocks in Indonesia went into a significant decline, and artisanal fishers as well as larger commercial concerns faced hunger and destitution or financial ruin if new fisheries were not accessed. Since the fisheries off northern Australia were relatively under fished, and were also relatively proximate to Indonesian waters, a huge increase in illegal fishing took place. During the summer of 2006–2007, it is estimated than in excess of 2000 boats were fishing illegally in Australian waters. In addition, many of these vessels travelled much further a field than had previously been the case, with many vessels being observed in areas such as the Gulf of Carpentaria in large numbers (*Torres News* 2007).

Responding to illegal fishing is clearly a maritime security issue. It involves the protection of natural resources subject to national jurisdiction from the over exploitation of others. Many States take the protection

Table 7.1 Selected EEZs and incomes

State	EEZ Area (km²)	GDP Per Capita (US$)
Federated States of Micronesia	2996419	2200
India	2305143	3100
Indonesia	6159032	4000
Kiribati	3441840	5300
Madagascar	1225259	1000
Marshall Islands	1990530	2200
Papua New Guinea	2402288	2300
Solomon Islands	1589477	2600

Sources: Sea Around Us Project (2010); CIA (2009).

of such resources extremely seriously, and it can be imagined that as such resources become more scarce and placed under greater pressure, the desire to protect will strengthen and become more widespread.

A difficulty in responding to illegal fishing in the exclusive economic zone (EEZ) is that the area over which many States have sovereign rights is too large to be effectively patrolled. Of States with an EEZ larger than 1 million square kilometres in area, eight have annual per capita GDP of less US$5300 (see Table 7.1).

It is also fair to observe that while India and Indonesia have some capacity to undertake patrol and enforcement within their EEZs, the other States in Table 7.1 have few assets that can be spared for such a purpose (Saunders 2008). Climate change will do nothing to assist these States improve their situation, and may lead to increasing pressure from illegal fishing within the EEZs.

3. MIGRATION

Another indirect impact upon the law of the sea could be a by-product of the environmental harm wrought by climate change and sea-level rise. It is not unreasonable to assume that if large areas of land become inundated, the inhabitants of these areas may be unable to remain without risking disease or famine. Higher sea levels may see the introduction of salt water through the tides into areas which are presently used for agriculture, rendering them infertile (IPCC 4th Report 2007, ch. 6). Similarly, sewage systems reliant on the expulsion of effluent into the oceans may be more prone to back up and failure

Even if populations are able to avoid direct inundation, they may still be substantially at risk from their increased vulnerability to storm and flooding in areas adjacent to those beset by the sea. While the most dire predictions see States such as Tuvalu and the Maldives evacuated, the displacement of potentially millions from areas such as coastal Bangladesh due to sea-level rise might also be expected (UNDP 2007).

In any circumstances where large numbers of people find their homes rendered unfit for habitation, or their access to uncontaminated food and water is severely restricted, experience suggests that large numbers of people will seek assistance elsewhere. Famine and natural disaster can be just as great a driver in the displacement of individuals as war and armed conflict. It might therefore be expected that if climate change deleteriously affects large populations in developing States, that many people from these populations will seek refuge in other parts of the world.

Individuals fleeing conflict, famine or poverty often make use of shipping to make their way abroad. This phenomenon is well demonstrated by events in the Mediterranean Sea where a large number of individuals from the southern, African, side of the Mediterranean seeking refuge in Spain and Italy reach these States by sea. This has proved to be a major maritime security concern for the enforcement authorities of these States. The United States Coast Guard is similarly extremely active in the interception of vessels carrying individuals bound for the United States without appropriate documentation. Over the past decade, Australia has faced sharply increasing numbers of individuals seeking to enter illegally by sea (Millbank 1999; Phillips and Spinks 2010). In the context of a deterioration of living conditions in some developing States as a result of climate change, it might be expected that instances like these proliferate and increase.

The international law implications arising from these movements of people are significant. In 2001, a Norwegian-registered container ship *MV Tampa* was denied entry into an Australian port as it sought to offload over 450 people who had been rescued at sea in waters south of Indonesia. Most of the individuals rescued by *Tampa* had paid for passage to Australia, a fact they forcefully made to the *Tampa*'s master, who after initially intending to disembark his passengers in Indonesia, subsequently made for the Australian territory of Christmas Island. Australia's refusal to take the rescuees was subsequently replicated by Indonesia and Singapore, as well as the flag State, Norway, potentially creating a humanitarian crisis aboard the ship, which was ill-equipped to sustain such a large complement for a long period of time. A crisis was ultimately avoided through the cooperation of Nauru, Papua New Guinea and New Zealand in agreeing to take the rescuees, on a temporary basis in the case of the former two States (Kaye 2002, pp. 59–81).

The *Tampa* case is by no means the only example, nor the most recent.

In the same region, the Australian Customs vessel *Oceanic Viking* was the subject of a stand-off between rescuees who were seeking to travel to Australia, and Customs officials who wished to disembark their charges in Indonesia. In this instance the vessel was able to discharge its human cargo in Indonesia, but only after the negotiation of arrangement apparently fast-tracking the acceptance of the rescuees in a number of developed States, including Australia (Thompson 2009).

At international law, the treatment of refugees claiming asylum is spelled out in the Convention relating to the Status of Refugees (Refugees Convention 1951) and its Protocol (Refugees Protocol 1967) which deal with the obligations upon State Parties in dealing with refugees arriving in its territory. A refugee is defined under Article 1 as a person with a well-founded fear of persecution for reasons of race, religion, nationality, membership of a particular social group or political opinion, who is outside their country of nationality, and is unwilling or unable to seek its help, and is unwilling to return. Most importantly in the context of incidents like those of the *Tampa* and *Oceanic Viking* is the obligation upon States under Article 33 of the Refugees Convention, which contains the *non-refoulement* principle:

> 1. No Contracting State shall expel or return (*refouler*) a refugee in any manner whatsoever to the frontiers of territories where his life or freedom would be threatened on account of his race, religion, nationality, membership of a particular social group or political opinion.

This provision restricts a State that has individuals with refugee-status from deporting those individuals to where they might be persecuted, and has the effect of restricting the similar expulsion of any person claiming such status at least until that claim has been reviewed. Accordingly, at international law, if individuals approach a State by sea, a contracting State like Australia could not deport them back to their country of origin, without breaching an important provision of the convention, and of customary international law.

However, it can be noted that where, as in the cases of *Tampa* and *Oceanic Viking*, the rescuees at no stage entered the State of arrival's territory, the standard that may be applied is one of seeking asylum through extra-territorial means. Shearer notes there is a clear difference in the application of principles of asylum between cases occurring within the territory of a State, and cases occurring extra-territorially (Shearer 1994). In the latter situation, the granting of asylum is typically more difficult and limited, and often involves a most direct threat to the life of the individual seeking asylum. However, it should be noted that these cases are principally concerned with entry to legations and embassies, and not to vessels floating about in a State's territorial sea. Certainly it is apparent that Australian policy was to avoid allowing individuals to reach its territory and to minimize the obligations owed. This is not dissimilar to the approaches of Italy and the United States facing similar boat arrivals bound towards their territory (Kaye 2002, pp. 59–81).

Climate changes raises the prospect of greatly increasing the movement of persons seeking a better life abroad. It is apparent from the examples above that developed States may resist the movement of 'climate refugees', and seek to avoid having individuals arrive within their territory, in order to minimize the duty they might owe. If the numbers of individuals seeking to move rises significantly, it might be expected that developed States will take as a restrictive a view as they may of their obligations, if for no other reason than the security implications in vetting increasing numbers of unexpected arrivals. Unregulated movements of large numbers of people may allow for greater freedom of movement for those engaged in terrorist activities.

4. FREEDOM OF NAVIGATION

The link between climate change and freedom of navigation may not be immediately apparent. While some of the more sensational predictions of climate change might include an increase in the number of major storm events around the world, it is not contended that these would reach a crescendo sufficient to detrimentally affect international shipping in anything other than the very short term. Rather, the impact upon navigation could be less direct but nonetheless very important.

Climate change has been increasingly linked in international literature to national and regional instability. The pressures wrought by a decrease in food production and the loss of possibly large areas to agriculture could bring other problems to the fore. At present, many developing States face internal instability because of extreme poverty and its consequences. It is no coincidence that of the five States with the lowest estimated per capita GDP, Somalia, Liberia, Burundi, the Congo and Zimbabwe, all five have faced or are facing substantial internal pressures or even civil war over the past decade (CIA 2009). It is not unreasonable to assume that if climate change ravages the economy of an already impoverished State, there is some risk that the State could descend into civil commotion.

This conclusion is not one which this chapter is alone in advancing. In a report written for the CNA Corporation by retired senior United States military commanders, including General Anthony Zinni, former US CENTCOM commander, it was stated:

> In the national and international security environment, climate change threatens to add new hostile and stressing factors. On the simplest level, it has the potential to create sustained natural and humanitarian disasters on a scale far beyond those we see today. The consequences will likely foster political instability where societal demands exceed the capacity of governments to cope.
> Climate change acts as a threat multiplier for instability in some of the most volatile regions of the world. Projected climate change would seriously exacerbate already marginal living standards in many Asian, African and Middle Eastern Nations, causing widespread political instability and the likelihood of failed states. (CNA Report 2007, p. 8)

A possible outcome for maritime security arising out of a failed State can be seen in the tremendous growth in Somali piracy in the past five years. Somalia is widely regarded as a failed State. Its government does not have effective control over large areas of the country which have been in open rebellion for decades. Many parts of Somalia have no effective law enforcement capability, and when combined with grinding poverty and easy access to cheap weapons, in this environment it was inevitable that illegal activity would dramatically rise. Given Somalia forms the southern littoral of the Gulf of Aden, the entrance to the Red Sea, through which the bulk of seaborne trade between Western Europe and the Middle East and southern Asia passes, it is not surprising that piracy flourishes.

The international community's reaction to Somali piracy demonstrates both the likely response to similar threats in the future, but also the limitations of such a response. International law treats piracy as an international crime, but usually only permits action to enforce the law against piratical acts in a manner consistent with State sovereignty. That is, a State may proceed to arrest a pirate provided it does so in areas subject to its own sovereignty or that are not subject to the sovereignty of any State. This limits enforcement against piracy by a State to its own territorial sea or to the high seas and EEZs of itself or other States. This is reflected in Article 105 of the Law of the Sea Convention (LOSC 1982):

> On the high seas, or in any other place outside the jurisdiction of any State, every State may seize a pirate ship or aircraft, or a ship or aircraft taken by piracy and under the control of pirates, and arrest the persons and seize the property on board. The courts of the State which carried out the seizure may decide upon the penalties to be imposed, and may also determine the action to be taken with regard to the ships, aircraft or property, subject to the rights of third parties acting in good faith.

For the purposes of the above Article, the Article 58(2) of the Law of the Sea Convention interprets the high seas to include the EEZ.

This restriction on enforcement actions against piratical acts created a potential difficulty in dealing with pirate attacks off the Horn of Africa, as Somalia lacked the ability to take any action to prevent attacks within its territorial sea. Operation of international law would mean that other States would not be in a position to apprehend pirates or take action against them, beyond what was necessary in self defence for an attacked vessel. Recognizing this deficiency, the United Nations Security Council adopted Resolutions 1816 (2008) and 1846 (2008). It is worth noting that even in this circumstance, there was some reluctance from the Security Council to authorize anti-piracy operations in the territorial sea, as there was a requirement of authorization for States to have authority from the Somali Transitional Federal Government (TFG), and for the resolutions to have limited periods of time in which they would operate. In addition, United Nations Security Council in resolution 1851 stated *inter alia*:

> 10. *Affirms* that the authorization provided in this resolution apply only with respect to the situation in Somalia and shall not affect the rights or obligations or responsibilities of Member States under international law, including any rights or obligations under UNCLOS, with respect to any other situation, and underscores in particular that this resolution shall not be considered as establishing customary international law, and *affirms further* that such authorizations have been provided only following the receipt of the 9 December 2008 letter conveying the consent of the TFG.

It is very clear that the Security Council did not want to create any State practice that might provide a basis for arguing that customary international law would accommodate interventions against pirates within another State's territorial sea.

In the wake of the Security Council's intervention, a large number of foreign naval vessels have been operating off the coast of Somalia, attempting to safeguard international shipping. Navy ships from over 30 States including the United States, India, China, the United Kingdom, France, Canada and Australia have all participated in these anti-piracy operations, with varying degrees of cooperation between the navies concerned (Azubuike 2009).

The above discussion would seem to indicate that international law, and international cooperation, were well equipped to deal with the consequences of Somali piracy, and therefore should be equally well equipped to respond to similar threats to maritime security in the future. While from a purely legal perspective this is true, reality has proven far more difficult to deal with.

Although a large number of foreign naval vessels have been operating off the coast of Somalia, very few have taken steps to exercise their jurisdiction to arrest pirates and bring them before a court. One reason ascribed to this apparent reluctance has been the possibility of having to bring alleged offenders and witnesses to the crimes back to their home States in order to try the pirates. The expense and difficulty this would entail is quite significant, and compounded by the risk both offenders and witnesses might seek asylum in the prosecuting State. In addition, as was observed in a different context in the preceding section, piracy, and the effort needed to combat it, may lead to people seeking entry into developed States. The prospect of escape from Somalia to a relatively wealthy developed State could see an increase in piratical activity rather than stamping it out. Efforts to prosecute individuals for piracy in jurisdictions less likely to be an attraction have met with mixed success at best (Askins 2010; Kraska and Wilson 2009).

The pirate activity off the coast of Somalia was not caused by climate change. It owes much to a breakdown in the rule of law in Somalia, grinding poverty, and the presence of foreign vessels in the waters nearby. These conditions have seen pirate activity increase in the past, as in the Straits of Malacca or in West Africa, although it might be observed that when the rule of law is reintroduced and effective enforcement of the law begins to occur, as was the case in Indonesia with the end of the fighting in Aceh in 2005, it can be expected that pirate activity will reduce (Kraska and Wilson 2009).

It is submitted, however, that climate change is relevant in the context of that already discussed above in the CNA Report. That is, climate change is a factor that may potentially exacerbate an already difficult situation (CNA Report 2007). Instead of a single failed State, in a world where climate change has seen significant environmental and therefore economic degradation, it might be expected that a number of States might fail. Poverty-stricken individuals, whose governments are unable to provide access to the necessities of life nor maintain law and order, may be driven to preying upon international shipping or offshore oil and gas installations. A reluctance to prosecute these people may see the practice of pirate raiding spread within communities and beyond to other failed or failing States. The practical outcome could be a world where the free movement of shipping around the globe could be drastically impeded by the activities of bands of pirates.

5. THE ARCTIC

If any region in the world is likely to face difficulties for maritime security arising directly from climate change it is the Arctic, although the nature of this security issue is of a very different nature to that posed in the paragraphs above. At present the Arctic Ocean remains inaccessible to normal shipping for much of the year, as it is covered with shifting ice. This ice has historically rendered the two principal Arctic sea routes – the Northwest Passage to the north of Canada and the US and Northern Sea Route to the north of Russia – closed to shipping, at least most commercial shipping. However with the sea ice of the Arctic reportedly in retreat, there is an increasing window each summer where passage through the two passages becomes feasible for large commercial vessels.

The Arctic Ocean potentially could provide links between the Pacific Ocean and the Atlantic Ocean via two routes. One route, the fabled North West Passage, passes through the maze of islands in northern Canada, while the other, the Northern Sea Route, passes along the northern Asian littoral. Both, if clear, offer the prospect of substantially shorter and faster transits than traditional routes through the Panama Canal, Cape Horn or the Indian Ocean. However both are challenging routes, passing through relatively confined waters in places that are poorly charted. In addition both routes pass through areas of high environmental sensitivity and both are remote from centres of population or places from which a rescue or environmental remediation could be staged (Butler 1978; Pharand 1984).

The coastal States along both these routes have traditionally been reluctant to open them to international navigation in a similar fashion to what is the norm elsewhere in the world. Since 1986, Canada has asserted territorial sea baselines around its Arctic Archipelago, effectively making the waters of the North West Passage internal waters (Order-in-Council 1985). This would mean that Canada would reserve to itself the right to determine if vessels could pass through the Passage. This certainly appears to be the case, given the ongoing dispute between Canada and the United States over passage rights through the Arctic Archipelago stretching back to the voyage of the *Manhattan* in 1968. A number of international vessels since the *Manhattan* have passed through the North West Passage, in spite of Canadian objection (Pharand 1984, pp. 47–53).

By comparison, Russia has typically taken a more restrictive view over the Northern Sea Route. Efforts by the US Coast Guard to pass through the Northern Sea Route in the 1960s were physically resisted by the then Soviet Union. Ice conditions along the Northern Sea Route meant that the only practical route required passage through the Vil'kitskii Straits, which Russia has sought to regulate. International passage through the Northern Sea Route has always been in company with a Russian icebreaker, with Russia of the view this has not rendered the Route open to international navigation, thereby attracting the transit passage regime under the Law of the Sea Convention (Butler 1978, pp.122–127).

The freeing of the Arctic from ice, at least through the summer, raises a number of security concerns for Russia and Canada. Both States have sought to substantially restrict international navigation through the waters of their Arctic littorals, and both have been aided in their objective by the appalling weather and sea conditions that dominate the region. However as conditions improve, both States' policies are likely to face increasing international pressure to change, particularly from China, Japan, the United States and the States of the European Union, who all see benefits in reducing travel times for sea traffic.

Another issue that may also raise a security concern has been the attitude of the Arctic States to their offshore jurisdiction in the Arctic. Russia, having lodged the first submission to the Commission on the Limits of the Continental Shelf, has taken a provocative position in respect of its continental shelf rights in the Arctic. The symbolic planting of a Russian flag on the continental shelf by a submarine on national television was a remarkable exercise in propaganda, but also raised the prospect of a confrontation over resources in the increasingly accessible Arctic Ocean. The Russian piece of theatre was the subject of a protest by Canada and the United States (BBC 2008).

6. CONCLUSION

Climate change and maritime security will remain key items on the agendas of States for the foreseeable future. States will face the increasing challenges. What is less clear is when and to what extent the two issues may coalesce over the coming decades. Certainly if some of the more dire predictions concerning climate change are correct, then the two issues will certainly become increasingly inter-connected as time goes by. As different parts of the Earth are increasingly affected by climate change, the impact upon large numbers of individuals will only increase, potentially placing greater strain on ocean resources, and setting ever growing numbers of people in motion, seeking a better life. This will ensure that issues in and around maritime security will continue to proliferate, in concert with climate change.

REFERENCES

Askins, Stephen (2010), *Piracy – A Review of 2009*, accessed 26th July 2012 at: http://www.incelaw.com/whatwedo/shipping/article/shipping-e-brief-february-2010/Piracy-a-review-of-2009.

Azubuike, Lawrence (2009), 'International regime against piracy', *Annual Survey of International and Comparative Law*, **15** (1).
BBC (2008), 'Arctic map shows dispute hotspots', accessed 26th July 2012 at: http://news.bbc.co.uk/2/hi/staging_site/in_depth/the_green_room/7543837.stm.
Butler, William (1978), *The Northeast Arctic Passage*, Alphen aan den Rijn: Sijthoff & Noordhoff.
CIA (2009), *World Factbook*, accessed 26th July 2012 at: https://www.cia.gov/library/publications/the-world-factbook/index.html.
CNA Report (2007), *National Security and the Threat of Climate Change*, CNA Corporation, accessed 26th July 2012 at: http://securityandclimate.cna.org/report/.
COP-15 (2009), United Nations Climate Change Conference, COP-15 Copenhagen, accessed 26th July 2012 at: http://unfccc.int/meetings/copenhagen_dec_2009/session/6262.php.
FAO (2007), *Yearbooks of Fishery Statistics: Summary Tables*, accessed 26th July 2012 at: ftp://ftp.fao.org/fi/stat/summary/default.htm.
Feely, Richard A., Christopher L. Sabine, Kitack Lee, Will Berelson, Joanie, Kleypas, Victoria J. Fabry and Frank J. Millero (2004), 'Impact of anthropogenic CO2 on the CaCO3 system in the oceans', *Science*, **305**, 362–366.
IPCC 4th Report (2007), *Contribution of Working Group II to the Fourth Assessment Report of the Intergovernmental Panel on Climate Change*, M.L. Parry, O.F. Canziani, J.P. Palutikof, P.J. van der Linden and C.E. Hanson (eds), Cambridge: Cambridge University Press.
ISPS Code (2004), Amendments to the Annex of the International Convention for the Safety of Life at Sea (SOLAS) 1974 contained in resolutions 1, 2, 6 and 7 of the Conference of Contracting Governments and Including the Ship and Port Facilities Security (ISPS) Code, London, 12 December 2002; entered into force 1 July 2004; [2004] ATS 29.
Kaye, Stuart (2002), 'Tampering with border protection: the legal and policy implications of the voyage of the MV *Tampa* for Australia's maritime border protection policy', in C. Rahman (ed.), *Protecting Australia's Borders*, Wollongong: Centre for Maritime Policy, pp. 59–81.
Kaye, Stuart (2005), 'The proliferation security initiative in the maritime domain', *Israeli Yearbook of Human Rights*, **35**, 205–229.
Kraska, James and Brian Wilson (2009), 'The pirates of the Gulf of Aden: the coalition is the strategy', *Stanford Journal of International Law*, **45**, 243.
Kyoto Protocol (1997), Kyoto Protocol to the United Nations Framework Convention on Climate Change, Kyoto, 11 December 1997; entered into force 16 February 2005; [2008] ATS 2.
LOSC (1982), United Nations Convention on the Law of the Sea, 10 December 1982; entered into force 16 November 1994; 1833 UNTS 397.
Millbank, Adrienne (1999), 'Boat people, illegal migration and asylum seekers: in perspective', *Current Issues Brief*, **13**, accessed 14th August 2012 at: http://www.aph.gov.au/About_Parliament/Parliamentary_Departments/Parliamentary_Library/Publications_Archive/CIB/cib9900/2000CIB13.
Order-in-Council (1985), Order-in-Council, Privy Council of Canada 1985-2739.
Ozone Convention (1985), Vienna Convention for the Protection of the Ozone Layer, Vienna 22 March 1985; entered into force 22 September 1988; [1988] ATS 26.
Pharand, Donat (1984), *The Northwest Passage: Arctic Straits*, Dordrecht: Martinus Nijhoff.

Phillips, Janet and Harriet Spinks (2010), 'Boat arrivals in Australia since 1976', *Australian Parliamentary Library Background Note*, accessed 26th July 2012 at: http://www.aph.gov.au/About_Parliament/Parliamentary_Departments/Parliamentary_Library/pubs/BN/2011-2012/BoatArrivals.

Rahman, Chris (2012), 'The implications of climate change for maritime security forces', in R. Warner and C. Schofield (eds), *Climate Change and the Oceans: Gauging the Legal and Policy Currents in the Asia Pacific and Beyond*, Cheltenham, UK and Northampton, MA, USA: Edward Elgar.

Refugees Convention (1951), Convention relating to the Status of Refugees, Geneva, 28 July 1951; entered into force 22 April 1954; [1954] ATS 5.

Refugees Protocol (1967), Protocol relating to the Status of Refugees, New York, 31 January 1967, entered into force 4 October 1967; [1973] ATS 37.

Saunders, Stephen (ed.) (2008), *Jane's Fighting Ships 2008–2009*, Coulsden: Jane's.

Sea Around Us Project (2010), accessed 26th July 2012 at: http://www.seaaroundus.org/eez/.

Shearer, I.A. (1994), *Starke's International Law*, Sydney: Butterworths.

SUA Convention (1988), Convention for the Suppression of Unlawful Acts Against the Safety of Maritime Navigation, Rome, 10 March 1988; entered into force 1 March 1992; [1993] ATS 10.

SUA Protocol (2005), 2005 Protocol to the Convention for the Suppression of Unlawful Acts Against the Safety of Maritime Navigation, London, 14 October 2005, yet to enter into force; [2005] ATNIF 30.

Thompson, Geoff (2009), 'Oceanic Viking standoff ends', accessed 26th July 2012 at: http://www.abc.net.au/news/stories/2009/11/17/2745239.htm.

Torres News (2007), '"Only several hundred" illegal boats', 18 May 2007, accessed 26th July 2012 at: http://www.illegal-fishing.info/item_single.php?item=news&item_id=1584&approach_id=25.

UNDP (2007), *Human Development Report 2007/2008*, accessed 26th July 2012 at: http://hdr.undp.org/en/media/HDR_20072008_EN_Complete.pdf.

UNFCC (1992), United Nations Framework Convention on Climate Change, New York, 9 May 1992; entered into force 21 March 1994; [1994] ATS 2.

Winberg Pia, Rubio Ana and Kirkendale Lisa (2012), 'Climate Change and Marine Living Resources' in R. Warner and C. Schofield (eds), *Climate Change and the Oceans: Gauging the Legal and Policy Currents in the Asia-Pacific and Beyond*, Cheltenham, UK and Northampton, MA, USA: Edward Elgar.

8. The implications of climate change for maritime security forces

Chris Rahman

Although a consensus on the reality of climate change now prevails, the details of the problem remain unsettled. In particular, the precise local and regional impacts of the global phenomenon of climate change are unknown. The security-related consequences of such impacts are thus inherently speculative. Nevertheless, it has become an increasingly important aspect of the national security policy deliberations of many States, including both civil and defence force capability planning, to consider the potential security implications of climate change. Despite the prevailing uncertainty over specific impacts and their security implications, long-range planning can be undertaken based on the most likely types of consequences relevant to future national and regional security environments. In this way, potential security problems generated either directly or indirectly by climate change processes, and the capabilities required to respond to them, can be assessed in a generic sense. This chapter addresses the climate change implications for maritime security forces. Here, 'maritime security forces' is used primarily to refer to navies, although coast guards and other civilian sea-going enforcement agencies are also discussed.

The chapter takes the following approach. First, it establishes the potential for climate change to create negative conditions for national or regional security. This discussion has been set within the context of the wider debate over the relationship between the environment and security. Second, the chapter briefly assesses the possible operational impacts of climate change for maritime security forces, in terms of the demand for their services resulting from the consequences of climate change, and the implications for their future force structure developments. Note that a discussion of the science of the possible effects of climate-change-induced modifications to oceanographic conditions upon the conduct of certain types of naval operations is beyond the scope of this chapter. Third, the challenges faced by maritime security forces in reducing their own greenhouse gas emissions are discussed in detail. This section relates to concerns regarding energy security and fuel costs, and necessarily focuses upon the organizations most advanced in tackling these rapidly emerging policy issues: the United States (US) Navy and the United Kingdom (UK) Ministry of Defence (MOD). Finally, the chapter concludes with an argument that, as with climate change itself, any implications for maritime security forces will be felt only gradually, making radical changes in force planning unnecessary. However, the effects of climate change will be distributed unevenly, and certain States may need to factor such implications into their planning sooner than others.

1. SECURITY IMPLICATIONS OF CLIMATE CHANGE

Climate change is accepted to have various and diverse potential security implications (Barnett and Adger 2007; CNA Corporation 2007; Dupont and Pearman 2006; German Advisory Council 2008; Pumphrey 2008). For example, the consequences of climate-change-induced, or at least climate-change-enhanced, damage to ecosystems and the natural environment generally is well recognized, and negative consequences are expected for human populations in terms of their economic, health and broader social well-being (Barnett and Adger 2009). Other anticipated outcomes of climate change include a loss of land for both economic sustainment and dwellings, and pressure on natural resources. In most cases, these impacts will not have direct security implications other than at the level of individuals or localized communities. However, when contemplating the implications for security forces, these so-called 'human security' factors, or the envisioning of security in terms of 'well-being', are not particularly useful (Rahman 2009). More salient are the downstream consequences of climate change, which may have significant effects on entire nations or regions.

Such consequences have the potential to create or contribute to realistic problems of disorder, on land and at sea. In the worst cases, one can even envisage genuinely strategic implications resulting from climate change-related conflict spilling across borders, and violent conflict between States. However, it is important to note that higher order impacts, such as the possibility for violent conflict and an unsettling of strategic security, will tend to be indirect, rather than operate as a direct consequence of climate change. Thus, the physical impacts of climate change are unlikely in and of themselves to create the types of conflict requiring military intervention, although the need for humanitarian assistance may be a more direct consequence. However, climate change impacts may generate negative political consequences by fuelling instability, disorder and inter-State competition, which may in turn have implications for the prospects for violence, whether internal, transnational or international. This argument mirrors the one commonly made for linkages between the environment and security. Such linkages are particularly important in those cases in which environmental degradation has been observed to affect security indirectly, and especially where violent conflict is involved. When considering such a relationship, environmental factors can be viewed as contributing to conflict situations only through the influence of a number of intervening variables: an independent impact by environmental factors is almost impossible to discern (Levy 1995). Moreover, to elucidate the issue further, climate change impacts that have implications for the internal stability of States and regional order are most likely to exacerbate existing societal and political pressures rather than independently create new ones. Only in the case of abrupt climate change (an extremely unlikely, but potentially catastrophic, scenario) would security implications be imposed independently.

In the Asia-Pacific region, likely localized consequences for the natural environment of global warming – or in some cases, at least, of global warming as a contributory or exacerbating factor – include: an increased frequency and severity of major storms, such as tropical cyclones and typhoons; increased incidences of other meteorological factors, such as drought or flooding; damage to ecosystems, both terrestrial and marine, and the food and other economic resources they contain; land degradation leading to the loss of arable land; and the slow, incremental loss of land through sea-level rise, and the related (and more probable) medium-term threats from storm surges and salt water inundation of fresh water tables (Intergovernmental Panel on Climate Change 2007; US National Intelligence Council 2009). The security implications of these localized consequences may include such factors as a spread of tropical diseases and other public health problems, and shortages of food and fresh water. For example, it is estimated that up to 100 million people in Southeast Asia alone rely upon marine resources directly or indirectly for food or as a source of livelihood (Williams 2007). Any damage to coral reefs – or any other element of the marine ecosystem, which is so vital for the health of the region's fish stocks – would have a major impact upon food security and local economies. In the Southeast Asian example, it is easily recognizable that a collapse in fish stocks would create real hardships and, potentially, political instability in these, and the most vulnerable, States.

Within States, such climate-change-influenced effects create potential for economic disruption and fiscal challenges for governments, internal migration, social unrest and political instability. These types of problems will be more severe for the relatively weak or small developing States of

Southeast Asia, South Asia and the southwest Pacific, many of which are already suffering from economic and governance shortcomings; which are in some cases exacerbated by demographic challenges and existing ethnic, religious and political tensions, and separatist movements. Potential disorder is also likely to be strongly represented at sea, for the Asia-Pacific region is largely maritime in character. Domestic instability and/or economic pressures are likely to spill over from the land environment onto the water, with increases in illicit activity such as illegal fishing, smuggling, illegal migration, and piracy and sea robbery. There is ample recent evidence to support this contention. For example, the Asian economic crisis of 1997–98 triggered an increase in piratical activities in and around the Indonesian archipelago (Mak, 2007); an anarchic internal environment provided the conditions for the Somali piracy that has been challenging the safety and security of shipping in the Gulf of Aden and adjoining parts of the Indian Ocean (Weir 2010); and an oil spill from the stricken Montara oil rig in the Timor Sea reputedly damaged fisheries and seaweed farms in West Timor, whereupon some of the affected fishermen turned to people smuggling, transporting illegal migrants transiting through Indonesia to Australia (ABC Radio Australia 2010).

There is real potential for the internal problems of one State to spill across into others. This is perhaps even more the case for instability at sea: an environment inherently transnational in nature. If climate-change-influenced challenges are added to the range of existing land and maritime territorial disputes, undelimited maritime boundaries and marine resource disputes, and deepening strategic competition between the region's major powers centred on China's rise, then the prospects for inter-State conflict are only likely to be enhanced in the long run. One can speculate that the inability to conclude an international consensus on measures to mitigate and combat climate change, and the pursuit of zero-sum games dominated by narrowly national interests to protect carbon-polluting habits, might also actually act as a spur to exacerbate international tensions.

1.1 Thinking about Climate Change and Security: the US Navy Example

In most respects, and certainly with respect to publicly announced policy and plans, the US Navy is the organization that has contributed most to thinking about the implications of climate-change issues for maritime security forces. The US Navy established Task Force Climate Change in May 2009 under the directorship of the Oceanographer of the Navy. It is tasked with assessing climate change implications for strategy, policy and plans, to include capability planning and environmental assessment. The initial focus is on the Arctic, where the US has direct territorial and jurisdictional interests, and where the climate is warming at a much faster pace than over the planet as a whole. Thus, the Arctic acts as a kind of living laboratory, which may presage the climate effects which will impact upon other regions in the decades ahead. This Arctic focus resulted in the promulgation of a first policy document, the *US Navy Arctic Roadmap*, released in October 2009, which establishes priorities and objectives for the Navy's planning related to the Arctic. The priority areas identified in the *US Navy Arctic Roadmap* for research, action and planning include: environmental data gathering and modelling; the potential for increased competition and conflict over access and natural resources; freedom of navigation; safety of navigation and search and rescue consequences of retreating sea ice; the need for Arctic maritime domain awareness (MDA); and assessment of the Navy's on-going ability to operate in the Arctic environment. The policy and strategy objectives will be operationalized by a forthcoming *Navy Arctic Strategic Implementation Plan*, while Navy capabilities and operational limitations will be assessed in a *Fleet Readiness Assessment* for Arctic operations.

A second document released in April 2010, the *US Navy Climate Change Roadmap*, addresses the naval planning implications of climate change beyond the Arctic. It is intended that both roadmaps will be reviewed and updated following the release of each *Quadrennial Defense Review*, the next version of which is due in 2014. The *Climate Change Roadmap* calls for the development of a *Global Climate Change Strategic Assessment* to identify the extent to which demand for certain types of naval tasks may increase because of climate-change impacts. Missions of particular interest include maritime security, humanitarian assistance and disaster relief, defence support of civil authorities, MDA, search and rescue, and strategic sealift. However, the roadmap's concerns stretch beyond just the possible implications for naval missions to include matters related to future force structure and infrastructure (US Navy 2010, p. 9). Regarding infrastructure, a particular focus is placed on climate-change impacts (such as sea-level rise, storm surges and erosion) on naval bases and other ports used to support US Navy operations, both in the US and internationally (US Navy 2010, pp. 12–13). For example, the low-lying Indian Ocean base of Diego Garcia (leased from the UK and important for the support of US operations in the Middle East and Afghanistan) is likely to be a prime candidate for such concerns.

Building on the *Arctic Roadmap*'s identification of the need for improved environmental assessment, the *Climate Change Roadmap* intends to develop improved regional and global environmental assessment capabilities to predict physical conditions in both the atmosphere and the oceans across varying time scales. These time scales will include those necessary for tactical purposes (hours and days), to periods stretching to weeks, months and even decades ahead. The *Climate Change Roadmap* identifies conditions of particular naval interest as including sea-level rise, the frequency and intensity of tropical storms, rainfall patterns and concentrations of greenhouse gases (US Navy 2010, pp. 16–19). This roadmap also identifies potential long-term impacts that could have important implications for the navy, including ocean acidification, geo-engineering and the possibility of abrupt climate change (US Navy 2010, p. 8). Importantly, it is intended that the assessments made by Task Force Climate Change will feed into other navy and defence policy documents, thus elevating the status of climate change as a potentially serious security issue for the US and its global interests.

2. CLIMATE CHANGE AND MARITIME SECURITY FORCE PLANNING

A useful way of thinking about future climate-change-generated demand for maritime security force responses is through the prism of the 'span of maritime tasks' typology commonly employed by navies of the British Commonwealth. This conception divides maritime operations into diplomatic (both benign and coercive), constabulary (from maintenance to enforcement) and military (combat and combat-related) functions (Booth [1977] 1979; Royal Australian Navy 2010, p. 100). The following discussion outlines the types of operations most likely to be generated or influenced by climate change using this typology, and the types of forces that will be required to conduct such operations.

The climate-change-related maritime role of greatest likelihood in the category of diplomatic operations is that of humanitarian assistance and disaster relief (HA/DR). With major adverse meteorological events widely predicted to be a feature of a global-warming-affected world, HA/DR operations by sea-based forces will provide an effective means of assisting battered coastal States. In a recent example, medium and heavy-lift helicopters from the US amphibious transport dock ship *USS Denver* were employed in disaster relief operations in Taiwan in August 2009 in the wake of Typhoon Morakot (Minnick 2009). Sea-based operations can be particularly important in situations in which terrestrial infrastructure has been destroyed, abandoned or is otherwise absent; or in cases in which there exist political sensitivities in the victim State about basing foreign forces on home soil. The latter issue can be illustrated by the ability of international, mainly US and Australian, forces to 'retreat' each night back to sea during the relief operations in Banda Aceh following the

Indian Ocean tsunami of Boxing Day 2004. This was necessary due to religious-cultural suspicions in the Islamic province, and in Indonesia more broadly, of foreign 'Christian' military forces basing themselves in the country (Elleman 2007, pp. 36–39). While not a weather-related disaster, this illustrates the capabilities of naval forces in responding to natural catastrophes on such a scale.

In responding to any demand for HA/DR missions from the sea, the size of the responding units is paramount. Generally speaking, the larger, more flexible and more capable the ship, and the larger its embarked rotorcraft, the greater is its capacity to respond to major crises, by being able to throughput more supplies or people in different mission profiles, as required. In particular, large, flat-deck amphibious ships and aircraft carriers represent important capabilities: the larger the flight deck, the larger the types of rotorcraft that can be employed, the greater the number of aircraft that can be operated and the higher the total number of missions that can be flown.

Another particularly useful attribute of many larger, more capable, amphibious ships and carriers is their integrated hospital facilities, often including operating theatres. In the Asia-Pacific region, the demand for such ships has been growing: in addition to the existing capabilities of the US, Indian, Japanese and Thai navies, the Korean Navy has recently added two helicopter carriers. Australia is also set to add two large amphibious helicopter assault ships of its own (Grove 2007, pp. 31–32; Fish 2010, pp. 22–27). Other navies, including that of China, are likely to follow suit in time. In the meantime, China, India, Indonesia and Singapore have all developed amphibious ship forces of the non-flat-deck varieties, all of which are still highly suitable for the HA/DR mission. Further, a sound command, control and communications capability, particularly the ability to interoperate with the forces of other nations in a cooperative or coalition environment, was demonstrated to be both difficult and vital during the tsunami relief operations (Elleman 2007). It is thus essential to build such interoperability in anticipation of multinational responses to future major HA/DR crises (Rahman 2008, pp. 35–49).

The capabilities described above are also useful for another relevant diplomatic function: evacuation operations. This maritime role may be particularly important in situations in which instability or a major public health emergency requires the rapid evacuation of one's own nationals or endangered civilians of the affected State. Indeed, there can be an overlap of various roles in some situations. In the example of the ravaging of New Orleans and other parts of the Gulf coast by Hurricane Katrina in August 2005, the US sea services (with assistance from foreign navies), led by the US Coast Guard, conducted massive HA/DR and evacuation operations, rescuing and evacuating approximately 33 000 people (US Naval Institute 2005).

The constabulary category of maritime tasks involves the gamut of 'good order at sea' roles (Till 2009), from search and rescue to environmental protection, maritime enforcement and border protection, and counter-piracy. All of these roles are likely to be in greater demand to varying extents because of the indirect effects of climate change-related impacts, particularly the combating of illegal activities such as illegal fishing, piracy and illegal seaborne migration (see Stuart Kaye's chapter in this volume).

The types of force structure required to undertake such constabulary operations will primarily involve patrol craft of varying sizes; from small patrol boats used for inshore work, larger patrol boats for operations throughout the exclusive economic zone (EEZ), to even larger offshore patrol vessels (OPVs), corvettes or frigates for more persistent operations in the EEZ and farther afield, and in heavier sea States. In the Polar Regions, the impact of climate change is already generating a demand for ships with strengthened hulls for constabulary operations in ice-affected waters (Fish et al., 2008, pp. 12–14; Hobson and Newell 2009, pp. 16–21). Airborne assets are also important, especially maritime patrol aircraft and, increasingly, unmanned aerial vehicles (UAVs) for maritime surveillance (Border Protection Command 2008, pp. 6–7). Indeed, it is well recognized that successful constabulary and maritime security operations require not just good surveillance capabilities but also increasingly sophisticated systems for MDA, which collect, integrate and analyse relevant maritime information for distribution to enforcement agencies (Rahman 2010, pp. 202–207). Developing effective MDA capabilities is a daunting challenge even for rich countries with extensive zones of maritime jurisdiction; but the task is much harder for developing States. Indeed, archipelagos such as Papua New Guinea, Indonesia and the Philippines, with their complex multi-island geography, make both MDA and the wider constabulary task particularly difficult. In most cases, these and other developing States of the Asia-Pacific region are desperately short of usable assets to support the constabulary function effectively.

Perhaps the least likely of the maritime functions to be required due to climate change is the military, or combat, role. Nonetheless, due to the factors discussed in the previous section, climate-change-influenced effects do have the potential to generate conflict, if only indirectly. This is particularly the case in those parts of the world in which political tensions already run high for a variety of other reasons, such as in South Asia, parts of Southeast Asia and Northeast Asia. While climate change is unlikely to be a significant factor in driving regional States to develop improved naval combat capabilities, nor will the probable need for greater numbers of constabulary and HA/DR assets lead them to de-emphasize combat roles. In many cases, coast guards or functionally similar agencies have been created as primary responders in the constabulary role, allowing navies to concentrate more on developing war fighting capabilities (Bateman 2003).

It is fortunate, then, that many naval platforms, in particular, are highly flexible and designed for operations across the span of maritime tasks. In general, the larger and more capable a platform is, the more flexible it will be. Large amphibious platforms represent a good case in point: designed primarily for amphibious assault, they are also perfect assets for HA/DR crisis response, as explained above; and HA/DR is often promoted at least as a secondary capability driver for the need for such assets. Surface combatants also tend to exhibit the flexibility to operate across the span of maritime tasks. In worst-case scenarios, in which climatic factors exacerbate existing conditions of burgeoning populations, other environmental strains, resource shortages and volatile political contexts, it is not too difficult to imagine a world in which competition for the resources needed to sustain populations, and attain or maintain strategic advantage over other polities, becomes the cause of increased conflict.

There also exist situations in which multiple types of operations from the span of maritime tasks, and stretching across the spectrum of conflict – from relatively benign circumstances at one end of the spectrum through to combat operations at the other – have been required, either sequentially or even simultaneously. As the Royal Australian Navy's *Australian Maritime Doctrine* publication explains, the span of maritime tasks typology should not be viewed prescriptively:

> . . . as the roles will rarely be exclusive and the tasks are representative rather than specific. In reality there will always be anomalies, not least because more than one of the roles and several of the tasks may be carried out simultaneously. . . . [T]he nature of any given operation can change rapidly. (Royal Australian Navy 2010, p. 99)

US and multinational operations in Somalia in the early 1990s provide a useful illustration of the potential for such operational complexity. Initially, in January 1991, US forces conducted an armed evacuation of foreign nationals. This was followed by US, and then multinational, humanitarian assistance in the form of famine relief, which subsequently developed into 'peace operations' in an attempt to stabilize the political situation, before eventually becoming a combat situation. The well-known setbacks that occurred during this period led, in 1994, to the withdrawal of US forces, and ultimately, in 1995, to a US-led withdrawal from this area of the UN contingent (Ohls 2009, pp. 2–4). Even though the Somali effort resulted in complete failure, almost all of these missions were nevertheless possible only by operations conducted 'from the sea'. Sea-basing was essential; and this example, as with the others described above, clearly demonstrates how important maritime capabilities will be in responding to the likely, or possible, security consequences of medium- to longer-term climate-change impacts.

3. REDUCING THE CLIMATE-CHANGE IMPACTS OF MARITIME SECURITY FORCES

In addition to undertaking long-range planning for the security implications of climate change impacts, defence organizations and other arms of governments, at least those in the developed world, are beginning to factor into their operational, environmental and budgetary planning considerations their own contributions to climate change. Indeed, it seems that defence organizations in the developed world will not be exempt from national attempts to reduce greenhouse gas emissions by the government sector. Much of that activity may be relatively benign for security forces, such as in improving the energy efficiency of buildings and motor vehicles, reducing emissions from defence estates and reducing non-essential travel (UK Ministry of Defence 2010a). However, reducing greenhouse gas emissions generated by security force operations or related operational activities, such as training and exercising, may be more problematic.

As a point of international law, the International Maritime Organization has no jurisdiction to regulate naval vessels and other government ships employed on non-commercial service. These vessels also enjoy the status of sovereign immunity from the marine environmental protection provisions (Part XII) of the Law of the Sea Convention (LOSC). However, States are urged to ensure that the operation of such vessels is consistent with those provisions, 'so far as is reasonable and practicable' (UN 1982, art. 236). Although hardly representing a licence to pollute, this caveat is sufficiently broad as to allow, in effect, a blanket exemption for ships and craft employed by national maritime security forces with respect to their environmental impacts. Any environmental constraints imposed on such government vessels is thus a matter for their flag State. In practice, many States, especially in the developed world, require a high level of compliance by their warships and other government vessels with national environmental regulations. For example, in the case of Australia, the Crown, and thus the Australian Defence Force, is bound by the provisions of the Environmental Protection and Biodiversity Conservation Act (1999). The

Royal Australian Navy (RAN) is obliged to comply with this Act (Kerr and Snushall 2005, pp. 162–163).

With respect to greenhouse gas emissions from ships, Australia is a party to the MARPOL Convention (International Maritime Organization 1978), and of most relevance, MARPOL Annex VI on the prevention of air pollution (International Maritime Organization 1978). The relevant Australian legislation is the Protection of the Sea (Prevention of Pollution from Ships) Act 1983 (Cth). However, as the Act gives effect to the Convention, which in turn grants sovereign immunity to warships, consistent with LOSC Article 236 (noted above), the Act does not apply to RAN vessels (Kerr and Snushall 2005, p. 162). In practice, though, Australia applies the MARPOL Convention to its naval vessels consistent with the Article 236 invocation for States to apply Part XII of the LOSC 'so far as is reasonable and practicable'. Nonetheless, the extent to which navies and other maritime security forces will be required to reduce their greenhouse gas emissions in the future will remain the exclusive preserve of the domestic regulatory provisions and attitudes of their respective governments.

In the US, the US Navy has instituted the Environmentally Sound Warship of the Twenty-First Century programme, which establishes high environmental compliance and protection standards for new classes of ships. The first ships to incorporate these environmental design benchmarks are the Military Sealift Command's *Lewis and Clark*-class Auxiliary Cargo Ammunition Ship (T-AKE) under way replenishment vessels (Markle and Gill 2006, pp. 89–91). The main propulsion units of the T-AKE ships are required to comply both with US Environmental Protection Agency and MARPOL Annex VI emissions standards for large marine diesels, and (in the case of auxiliary diesels) MARPOL Annex VI criteria (Markle and Gill 2006, p. 100). One of the initial environmental design stipulations was the prohibition of steam propulsion, to reduce total greenhouse gas emissions (Markle and Gill 2006, p. 104). Although the T-AKE ships will increase emissions of nitrogen oxides and hydrocarbons compared to legacy platforms due to the characteristics of diesel engines, total engine emissions will be only one-third that of legacy ships. In fact, despite operational employment plans for the T-AKE ships to increase the days spent under way by 34 per cent compared to earlier classes, total emissions reductions of the main greenhouse gas, carbon dioxide, are in the order of 53 per cent (Markle and Gill 2006, p. 107).

The overall contribution of the defence sector to national greenhouse gas emissions in most countries tends to be relatively small as a proportion of overall emissions, although total emissions can still be significant when divided by sector. This is particularly the case for those generated by military aviation and, to a lesser extent, by naval vessels. In the UK, MOD activities generate approximately 1 per cent of total national greenhouse gas emissions. Two-thirds of those emissions are generated by the 'operational energy use' of military ships, aircraft, vehicles and generators (UK Ministry of Defence 2010a, p. 6, n. 5), which has been defined as 'the energy used to deliver propulsive power and electricity in generation, sustainment and recuperation of fighting power' (UK Ministry of Defence 2010a, p. 6, n. 5). The Royal Navy's share of MOD emissions is believed to be around 19 per cent (Bailey 2009, p. 69).

In the US, emissions by the Department of Defense (DoD) have not been consistently tracked. In 1996, the Department was responsible for around 2.3 per cent of national greenhouse gas emissions (Office of the Under Secretary of Defense for Acquisition and Technology 2001). In terms of energy use, according to the Secretary of the Navy, in 2009 the DoD was responsible for approximately two per cent of US total energy use, and over 90 per cent of the energy consumed by the federal government (Mabus 2009). There is likely to be a relatively close correlation between DoD energy use and its proportion of national emissions, given the prevalence of fossil fuels in its energy consumption mix. Of that DoD energy consumption, in the 2007 fiscal year, three-quarters was accounted for by ships, aircraft and vehicles: including 53 per cent by jet fuel (all services) and a further 11 per cent by marine diesel fuel (US DoD 2008, p. 4, fig. 2). In terms of emissions, it has been estimated that 'all DoD marine sources' contributed only about one-thousandth of total US carbon dioxide emissions, and that the US Navy surface fleet generated around six million tons of carbon dioxide annually in the early years of the century (Breslin and Wang 2004, pp. 29–30).

The difficulty in reducing greenhouse gas emissions generated by operations is due in no small part to their highly contingent nature, with conflicts, external crises and demand for humanitarian assistance, or border protection challenges arising at short notice, often unexpectedly. Therefore, even establishing a meaningful 'baseline' from which to target reductions may be impossible for defence and other security forces. Even an annualized rate averaged over a number of years would almost inevitably be adversely affected by contingency. Nevertheless, the UK Ministry of Defence has tasked itself with establishing such as baseline, setting objectives to reduce emissions generated by the operational use of defence force equipment. Although acknowledging the challenges involved in achieving this objective while 'at least maintaining or preferably improving operational capability' (UK Ministry of Defence 2010a, pp. 6–7), the MOD has set deadlines of 31 March 2011 to establish baselines for equipment emissions, and 31 March 2012 to set targets for reducing those emissions

(UK Ministry of Defence 2010b, p. 6). It should be noted that the British armed forces have already attained their proportion of emissions reductions required by Kyoto Protocol 2012 targets for the UK, and have met the stricter European Union reduction targets required for 2020, relative to 1990 national baseline emissions. This is owing to the large-scale force structure reductions that have taken place since the end of the Cold War; equating to a 36 per cent reduction in Royal Navy force structure between 1990 and 2006 (Bailey 2009, p. 69). However, meeting national emissions reduction targets for 2050 would require a further 21 per cent reduction from 2006 baseline levels if the MOD were to deliver reductions proportionate to its contribution to total national emissions (Bailey 2009, p. 69). Further drastic force structure reductions announced in October 2010 as part of stringent UK budget cuts will no doubt assist in meeting those targets (HM Government 2010, pp. 21–23), even if operational capabilities will be inadvisably curtailed in the short term.

A similarly significant reduction is likely to have occurred with respect to the US Navy, given its greatly reduced force structure since the end of the Cold War (and notwithstanding the frequent interruption of major gas-guzzling contingencies). However, the same trend cannot be identified in the Asia-Pacific region, where maritime security forces of all kinds have been in a growth phase due to the expanding maritime interests of many Asian States and a concomitantly increased demand for offshore constabulary capabilities. This latter demand has been further driven in some cases by post-9/11 maritime security concerns, as well as deepening strategic competition, which has thus far been most acutely apparent at sea.

Taking a different tack to that of the UK, the US government has set a requirement for all federal agencies to increase energy efficiency and 'measure, report, and reduce greenhouse gas emissions from direct and indirect activities' relative to 2008 baselines. However, provision is made for the exclusion of emissions generated by the operational use of military vessels, vehicles and aircraft, and for the purposes of law enforcement and emergency response (The President of the United States of America 2009, s1, 2(a) and 19(h)). Despite the potential for such operational exemptions, including by US maritime security forces, and as outlined in detail below, the US Navy has taken upon itself the challenge of significantly reducing its greenhouse gas emissions and leading the national response to the security challenges posed by a warming climate.

However, given the fundamental nature of defence and security as a primary function of government, it is highly unlikely that any responsible government would allow such targets to undermine necessary operational capabilities in the pursuit of other, less fundamental objectives, such as reducing greenhouse gas emissions. Any government action on emissions that materially diminished defence force capabilities would ultimately be viewed as self-defeating. This situation could well be exacerbated if many of the consequences of global warming with concomitant security implications, as noted in the preceding sections, require an operational response from maritime security forces. This means that climate change itself, over time, may necessitate a scaling-up of operations by both military and civilian security forces.

Further, most significant sea-going and maritime aviation platforms used by security forces have long life spans, typically around 25 years, but sometimes stretching to 40 or even 50 years. This makes it difficult, and potentially very expensive, to engineer improved environmental standards, such as for greenhouse gas emission reductions, into an existing, often quite aged, force structure. In the British example, the difficulty of adapting current capabilities is well understood. Yet the *MOD Climate Change Strategy 2010* (UK Ministry of Defence, 2010a, p. 7) also stresses the need to ensure that future capabilities are fully adapted for lower emissions and other requirements of a climate-change-affected world.

3.1 Climate Change and Energy Security

There is a potential upside to this seemingly irreconcilable conundrum of operational versus environmental concerns. That upside resides in the close interrelationship between considerations of climate change and those of energy security, particularly as the latter relates to energy efficiency goals. Energy security has been a longstanding theme among the vital national interest concerns of most States. Its prevalence in national security thinking has waxed and waned, depending upon the energy context of the day: rising in times of Middle East crisis and during the oil shocks of the 1970s and 1980s, for example, and diminishing in times of assumed plenty and low world energy prices. In the first decade of the twenty-first century, a number of factors have coalesced to focus attention on energy security issues, particularly related to oil. Oil is the most strategically important of the energy sources due to the near monopoly of petroleum fuels in powering sea, air and most land transportation modes, including those used for military purposes.

The threats to energy security have included greatly increased demand from the large emerging economies such as China and India, war and on-going instability in the Persian Gulf region, uncertainty over the extent of remaining oil reserves in Saudi Arabia and other major Gulf producers, and the use of oil securities as tradable financial products by international financial markets. All of these potential threats influence the long-term trend of upwards pressure on energy prices. In addition, in some countries, measures to reduce carbon emissions also affect already rising energy prices. Steep price rises have been particularly evident since 2005. As an indication of the magnitude of the problem, the US Air Force reduced its consumption of aviation fuel between the 2003 and 2007 fiscal years by 17 per cent. However, the cost of fuel consumed during that period increased by 115 per cent (Billings et al., 2008, p. 23). The US Navy's energy costs have almost tripled since 1991 (US Naval Sea Systems Command 2010).

For navies and other maritime security forces, ensuring consistent access to adequate fuel supplies and cost mitigation have been the two overriding drivers of their respective fuel efficiency measures. It may be serendipitous then that improved fuel efficiency provides consequential, near proportionate, reductions in greenhouse gas emissions. Fuel efficiency is not a new issue for navies. The US Navy, for example, has had an active ship and aircraft fuel efficiency programme in constant operation since 1977, albeit as a relatively low priority and at low levels of funding (Office of the Under Secretary of Defense for Acquisition and Technology 2001, p. 50). Nonetheless, their Energy Conservation programme is reputed to have saved $157 million in fuel costs in 2008 alone (Billings et al., 2008, p. 23). Increasingly, fuel efficiency is being viewed as an important aspect of operational effectiveness, in large part due to a realization of the very great logistics and planning burdens, operational constraints and financial costs imposed by the need to maintain long, expensive and sometimes vulnerable petroleum supply chains. This reality has been highlighted by the on-going wars in Iraq and Afghanistan, as well as maritime operations in the Arabian Sea and the Gulf of Aden. However, it should be noted that if fuel savings result in ships being able to remain at sea longer – that is, by consuming the same amount of fuel, but increasing the time able to be spent under way – there will be no positive impact on greenhouse gas emissions.

The focus on reducing the high level of fossil fuel dependence across the services and reducing related costs and vulnerabilities over time has led to an emphasis on calculating the true cost of fuel consumed by military platforms. The US Naval Sea Systems Command (NAVSEA), for example, calculated the actual cost of fuel consumed by the fleet. When costs associated with transportation, storage and in-theatre delivery (including fleet oiler costs) were included, another 30 to 90 per cent was added to the 'standard' price of fuel (Office of the Under Secretary of Defense for Acquisition and Technology 2001, p. 20). Further, the in-theatre protection of oil supply logistics trains, depots and port facilities increases the cost again in conflict zones. This metric has become known as the 'fully burdened cost of fuel' in US parlance, and is now the required standard when calculating fuel costs throughout the US armed forces, as mandated by the 2009 National Defense Authorization Act (US DoD 2010, p. 87;

Tiron, p. 30). The UK has also adopted this concept for assessing the total cost of energy used for both training and operations, using the term 'fully burdened cost of energy'. This concept is to be employed when establishing greenhouse gas emissions reduction targets (UK Ministry of Defence 2010b, p. 6).

In the US, the current Secretary of the Navy, Ray Mabus, has prioritized energy reform as one of the three focus areas for his tenure (Mabus 2009, p. 1). In so doing, he has explicitly linked energy reform to carbon emissions. Although the primary rationale for embarking upon this particular reform is energy security, Mabus is nevertheless forthright in recognizing the contribution to climate change of military-generated emissions; stating that 'the Navy and Marine Corps have an obligation to do something now about our impact on the environment' (Mabus 2009, p. 3), and that 'it is an environmental necessity to begin to do some things to cut down the carbon footprint' of the navy (Mabus, cited in Tiron 2010, p. 31). Invoking great US ventures and technological advances of the past, Mabus committed the Department of the Navy in October 2009 to taking a leading role in developing and utilizing new energy efficiency technologies and operating procedures by establishing five energy targets to be implemented over the next decade.

First, an assessment of lifetime energy costs and the fully burdened cost of fuel used to power both systems and buildings will become a mandatory factor when evaluating the award of contracts. The energy efficiency and footprint of contractors will also be taken into account. Second, the Navy will establish a so-called 'Green Strike Group' by 2012. Invoking the voyage of Theodore Roosevelt's Great White Fleet of 1907–08, it is intended that the flotilla will deploy as a 'Great Green Fleet' by 2016, 'composed of nuclear ships, surface combatants equipped with hybrid electric alternative power systems running bio-fuel, and aircraft flying only bio-fuels'. Third, petroleum use by Department road vehicles is to be halved by 2015. Fourth, alternative energy sources are to account for at least half of the Department's energy use from shore-based installations by 2020. Finally, fifth, alternative energy is to account for half of the Department's total energy consumption by 2020, a total inclusive of fuel consumed by Navy and Marine Corps ships, aircraft, tanks and other military vehicles (Mabus 2009, pp. 7–9).

It is impossible to predict whether such ambitious targets will be met, or even whether they are actually achievable. However, they do represent strong evidence of intent. The seriousness with which the US Navy approaches these issues is also reflected by the establishment of Task Force Energy, which will produce a Navy energy strategy detailing the organization's climate change mitigation measures (US Navy Climate

Change Roadmap, p. 3). Task Force Energy is intimately connected to Task Force Climate Change, addressed earlier in the chapter. In combination, the two task forces place the US Navy at the forefront of efforts by security forces globally to deal with all aspects of climate change.

3.2 Solutions

There are a number of actual or potential solutions available to maritime security forces to improve fuel efficiency and thus reduce their greenhouse gas emissions. The following section addresses four types of solutions being implemented or under consideration; namely, changes to operational practices; alternative fuels; engineering and technological solutions capable of being retrofitted to existing platforms; and new and emerging technologies that could be integrated into future platforms.

3.2.1 Changes to operational practices

Relatively simple fuel-saving modifications can be made to the manner in which maritime security force vessels conduct operations, especially in operationally benign contexts. First, greater use can be made of improved voyage planning. By so doing, optimal voyage routes can be employed to reduce distances sailed from point A to point B, thus ensuring the lowest possible fuel consumption. Another recommended operational strategy is for ships to operate a single engine at its peak efficiency, rather than multiple power plants at reduced efficiency (or other variations on this theme) when undertaking slow-speed operations, such as non-urgent transits and patrolling. The use of such strategies could reduce fuel consumption by as much as 25 per cent (Breslin and Wang 2004, p. 31; Office of the Under Secretary of Defense for Acquisition, Technology and Logistics 2008, p. 32). They have been developed in the US by the NAVSEA Ship Energy Conservation Assistance Program (SECAP). This programme in turn informs the Optimum Ship Transit Program, which utilizes SECAP data to determine optimum transit speeds with respect to the impact of speed upon fuel efficiency (Breslin and Wang 2004, p. 31; Office of the Under Secretary of Defense for Acquisition, Technology and Logistics 2008, p. 32).

A number of operational procedures and solutions are also being considered by the US DoD to reduce aircraft fuel use. These solutions are relevant to all aircraft used in the maritime environment, ranging from the increased employment of simulators for training, to more efficient route planning and better mission planning to minimize the need to dump fuel. Other operational procedures restricted to larger, more capable navies – if not necessarily unique to US maritime forces – include the reduced use of afterburners in naval combat aircraft and reductions in the need for in-flight refuelling (Office of the Under Secretary of Defense for Acquisition, Technology and Logistics 2008, p. 32). Improved computer-based flight planning systems are estimated to reduce fuel use by three to six per cent (Office of the Under Secretary of Defense for Acquisition and Technology 2009, p. 54). The increased use of simulators and other techniques could also allow an increasing proportion of training and exercising, within the limits of practicability, to be conducted without having to put to sea (or air).

Unmanned (robotic) vehicles are already in widespread use in the three now traditional geographical environments, and offer significant advances both in reduced fuel use and in persistence, compared to the manned alternatives (Office of the Under Secretary of Defense for Acquisition, Technology and Logistics 2008, pp. 48–49). In the maritime environment, UAVs (Pengelley 2009, pp. 12–17), unmanned underwater vehicles (UUVs) (Funnell 2009, pp. 26–31) and unmanned surface vessels (Scott 2008, pp. 16–22) are all currently in service, and their development and employment is expected to grow rapidly worldwide. The robotics revolution, although already widespread and rapidly advancing, remains in its infancy (Singer 2009). For maritime security forces, UAVs used as airborne sensors – that is, for intelligence, surveillance and reconnaissance (ISR) to identify and track vessels, and thus potentially illegal or threatening behaviour at sea – are being integrated into complex MDA systems, and will increasingly be adopted both by navies and civilian enforcement agencies. The increasing integration of unmanned platforms into current operational plans and systems, and the replacement of existing, manned systems, with newly available unmanned ones, promises significant fuel efficiencies and emissions reductions over time. It has been speculated upon by one Royal Navy officer, for example, that the increasing integration of UUVs with the surface fleet for mine hunting operations could eventually lead to an entire type of ship – that is, the mine hunter – being rendered redundant and thus able to be deleted from naval force structures (Bailey 2009, p. 71).

3.2.2. Alternative fuels

The US Air Force has been pursuing new aviation fuel blends using both bio-fuels and synthetic fuels for some time, with potential benefits for naval and other maritime users. Synthetic fuel alternatives include coal-to-liquid processes. However, it will be important to avoid processes that 'exceed the carbon footprint of petroleum' (Office of the Under Secretary of Defense for Acquisition, Technology and Logistics 2008, p. 22). Indeed, the 2007 Energy Independence and Security Act precludes US federal agencies from procuring such fuels (Billings et al. 2008, p. 23). In addition to synthetic fuel production, the Naval Research Laboratory is pursuing research into a number of other alternative energy sources, including fuel cells and advanced batteries, the use of superconductors and the catalytic conversion of hydrogen derived from seawater (Tiron, p. 32). Ultimately, it may thus be possible for ships to produce at least part of the energy they consume while at sea directly from the sea and with minimal carbon dioxide emissions. It may also be possible to convert the thermal energy potential of tropical waters into electricity and to exploit the electricity-creating, carbon dioxide-consuming properties of certain marine micro-organisms: a form of energy that has already been developed to power small acoustic hydrophones moored to the seabed to monitor shipping (Tiron, p. 32).

Experimentation has also been taking place using a variety of bio-fuels. The US Navy has successfully test flown an F/A-18 Hornet combat aircraft on a 50/50 blend of regular naval JP-5 jet fuel and bio-fuel derived from camelina (Hsu 2010), a plant related to canola that can be crop-rotated with wheat (Biello 2009). It is planned that this fuel will power a so-called 'Green Hornet' as part of the Great Green Fleet plan described earlier, with the expectation that the fuel will also be trialled in ships (Reuters 2010; Mabus 2009, p. 5). The US Navy is also experimenting with algae-based bio-fuels and has already contracted for experimental volumes from a Californian company, Solazyme. This contract comprises 1500 gallons of an algae-based version of JP-5, Solajet HRJ-5, which reputedly produces 85 per cent fewer greenhouse gas emissions than the conventional naval jet fuel, and 20 000 gallons of an algae-based version of F-76 marine distillate fuel (Hsu 2010). Although production capacities are limited and costs are high, the attitude of Secretary Mabus is that 'If the Navy comes knocking, they [that is, industry] will build it. The price will come down and the infrastructure will be created' (Mabus quoted in Rosenthal 2010). It is clear that Mabus is committed to a long-term future that will be far less petroleum and carbon intensive, and that he has demonstrated a willingness to shoulder the burden of research and development costs, which will ultimately benefit other maritime users, including those in the civilian economy.

3.2.3. Solutions applicable to current vessels

There exists a range of solutions able to be applied or retrofitted to the vessels of existing maritime security force structures. A 2001 US Defense Department study outlined five basic categories of improvements that could be applied to existing vessels involving the following engineering elements: propulsion and auxiliary power systems; ship 'hotel' electricity loads; auxiliary systems; hydrodynamic improvements to hulls; and anti-fouling hull coatings. In the first category, a number of mechanical and electronic improvements are being pursued across all non-nuclear engine types. For example, the integration of modern fuel injection systems and better management of low-load operations for current diesel engines are estimated to reduce fuel consumption by up to five and 14 per cent respectively. The application of this range of technologies across the gas turbines, diesels and boiler plants currently in use is estimated to result in fuel savings in the range of 3–8 per cent (Office of the Under Secretary of Defense for Acquisition and Technology 2001, p. 51).

One new propulsion technology development with the potential for use on many existing vessels is the hybrid electric-drive system. The system employs a hybrid electrical-mechanical power arrangement that allows one engine to be shut down at low operating speeds, thus improving fuel efficiency. The system will be trialled on the DDG-51 *Arleigh Burke*-class guided missile destroyer, *USS Truxton*, in 2012, and could be retrofitted to existing ships 'without extensive modifications' (Walsh 2010, p. 88). Twelve DDG-51 destroyers are due to receive the system in the near term, with estimated annual fuel savings of almost $1 million per ship. The first ship to be equipped with a similar type of technology is the *USS Makin Island*, a large flat-deck amphibious ship. During its 2009 maiden voyage alone, on which it circumnavigated South America, it is claimed that use of its hybrid auxiliary propulsion system saved almost $2 million in fuel costs. Based on 2009 fuel prices, NAVSEA has estimated that the ship will save $250 million in fuel costs alone over its lifetime (Mabus 2009, p. 6). That clearly also equates to large reductions in greenhouse gas emissions. Complementing this, new engine component technologies to be retrofitted to existing aircraft systems are also being pursued. In the case of the F/A-18 combat aircraft used by both the US Navy and US Marine Corps, such technologies are expected to provide a 5 per cent improvement in fuel efficiency (Office of the Under Secretary of Defense for Acquisition and Technology 2001, p. 54).

Reductions in electricity consumed by power sensor and weapon systems is unlikely given that the power demands of modern electronics, radars and, in the mid-term future, laser and other advanced electrically powered weaponry types (Scott 2007, pp. 34–39, 2009, pp. 8–10) will almost certainly grow. However, there are other opportunities to reduce the power consumption of, for example, lighting and water desalination plants. It has been estimated that perhaps as much as 30 per cent of the fuel consumed by the US Navy for non-aviation purposes is used to generate ship hotel electricity loads. The application in ships of the best practice energy efficiency measures, designed for and used in, buildings (such as efficient forms of lighting and lighting controls) are likely to be relatively easy to retrofit and may reduce hotel loads by at least 20 per cent. Meanwhile, mechanical improvements to auxiliary systems (such as a ship's air conditioning) may result in savings of another 3–6 per cent (Office of the Under Secretary of Defense for Acquisition and Technology 2001, p. 53). A 2008 Defense Science Board Task Force on DoD Energy Strategy noted that not only are long-term operational benefits made possible by hotel load efficiency gains, but also the returns on energy efficiency investments are now far greater than at the time of the initial Defense Science Board report on fuel burden reductions in 2001. This is due to the significant rise in world fuel prices (Office of the Under Secretary of Defense for Acquisition, Technology and Logistics 2008, p. 47).

In the fourth category, the retrofitting of hydrodynamic hull improvements such as stern flaps, bow bulbs and bow fins are being pursued for different categories of ship, and are estimated to improve fuel efficiency by 3–8 per cent (Office of the Under Secretary of Defense for Acquisition and Technology 2001, p. 52). In the case of bow bulbs, it may be possible, for example, to integrate the sonar domes of some warships with a bulbous bow that allows for reduced resistance through the water. US Navy experiments suggest that were such a system to be integrated into the design of the DDG-51 destroyers, the mainstay of the US surface combatant fleet, power reductions of 5.7 per cent could be achieved, representing a 3.9 per cent reduction in fuel consumption (Breslin and Wang 2004, p. 34). Stern flaps are another means by which hydrodynamic drag can be reduced, and these are being retrofitted to all current classes of US Navy surface combatant and some classes of amphibious ships, and may eventually be applied across the entire fleet. They are relatively cheap and easy to install, and pay for themselves in fuel savings within one to two years of installation (Office of the Under Secretary of Defense for Acquisition and Technology 2001, pp. 52, 55), with fuel consumption reductions ranging between four to eight per cent (Breslin and Wang 2004, p. 32).

Finally, in the fifth category, the use of advanced anti-fouling hull coatings can significantly reduce fuel consumption by increasing a ship's hydrodynamic efficiency. Fouling from biofilm alone can increase a vessel's drag through the water by up to 20 per cent, while barnacles and other marine life growth can increase drag by over 60 per cent (US Office of Naval Research 2009c). It has been estimated by the US Navy's Office of Naval Research (ONR) that, over a period of six months, a ship can acquire 150 kilograms of barnacle weight per square metre. Such bio-fouling rates can reduce speed by up to 10 per cent, which in turn requires fuel consumption to increase by up to 40 per cent to counter the effects of the accumulated drag (US Office of Naval Research 2009a). In dollar terms, this represents an additional annual cost for the US Navy (in extra fuel and increased maintenance) of approximately $1 billion (US Office of Naval Research 2009c). Bio-fouling is a particular problem for many classes of naval ships, which often tend to be dry-docked quite infrequently, thus allowing for a greater accumulation of fouling than occurs on commercial ships between limited opportunities for hull cleaning and repainting (Breslin and Wang 2004, p. 34). The issue is compounded by an International Maritime Organization ban on highly toxic tributyltin-based hull coatings, and moves to shun commonly used copper-based coatings due to the negative environmental effects of copper's biocide properties (Breslin and Wang 2004, pp. 34–35). Although the legal principle of sovereign immunity once again applies, many navies have been actively pursuing new anti-fouling technologies to overcome the problems of toxicity, fuel consumption and cost and carbon emissions; and to satisfy the need for longer lasting naval vessel coatings with extended effective life spans to stretch between scheduled dry-dock periods. The types of solutions being pursued by navies and a number of hull coating manufacturers include non-toxic polymers and resins, such as easy release silicone coatings and coatings incorporating quickly biodegradable biocides (Breslin and Wang 2004, p. 34).

3.2.4 New and emerging technologies

One technology that has been under development since the mid-1980s, involving collaboration between the US, British and French navies, is the intercooled recuperated (ICR) gas turbine engine. This system employs water to cool intake gases between the compressor sections and a recuperator to recover heat energy from exhaust gases, which is then used to pre-heat air used for combustion. The Rolls-Royce WR-21 ICR engine has many advantages over older types, including reduced maintenance, through-life costs and thermal signature. Most importantly for our climate change focus, however, is that the ICR cycle greatly reduces fuel consumption, and thus greenhouse gas emissions. Engine-maker Rolls-Royce (n.d.) claims that the WR-21 reduces fuel use by an impressive 25–27 per cent compared to current generation marine gas turbines, while the Royal Navy's Warship Support Agency expects fuel savings of up to 30 per cent when used in large warships (English 2003, p. 5). The Royal Navy has become the first maritime force to adopt the WR-21, selecting it for its new Type 45 *Daring*-class guided missile destroyers. The engine is suitable for use in a number of propulsion configurations and across a wide range of surface combatants, from corvettes and frigates through larger destroyers and cruisers, and in aircraft carriers (English 2003, pp. 4–5). The WR-21's high efficiency in different modes of operation also enables it to be used both at optimal cruising speeds for greatest fuel efficiency and as a 'boost' engine when operational requirements demand bursts of higher speed (English 2003, p. 3). This ability allows a single efficient engine type to be used for both operational modes, thereby reducing the number of engines designed into any particular ship.

The medium-term vision for the US Navy (and many other modern forces) is to move to full electric drive for the next generation of ships; at least for classes of larger vessels in the 7000 to 8000 ton full-load displacement range and above (Rosamond 2008, pp. 20–25). Indeed, the Royal Navy's Type 45 destroyers use this type of integrated electric propulsion (IEP) system. In its initial sea trials in 2007, with its two WR-21 ICR gas turbines generating the power for the IEP system, *HMS Daring* covered over 4000 nautical miles with an average daily fuel usage only one-quarter that of the classes of ships it will replace. Further, it has been estimated that the through-life costs for the Type 45 will be 45 per cent lower than that of its predecessor class of destroyer (Rosamond 2008, p. 22). Electric drive uses an integrated power system that utilizes a ship's engines to provide power for propulsion, weapons and sensors, and other ship services via an electric motor, eliminating many heavy and potentially vulnerable mechanical components, and providing greater flexibility in engine employment. This ensures that peak efficiency can be maintained across different operating profiles. Further ONR research into electric-drive systems is developing technologies to allow excess power to be stored rather than dissipated (Walsh 2010, p. 88). While the hybrid electric-drive system noted earlier is already in service, fully integrated electric-drive systems seem almost certain to feature in most large, next generation Western warships, offering significant fuel, and thus emissions, savings, in the order of 15 to 19 per cent (Breslin and Wang 2004, p. 36).

Full-cell technology that uses the electro-chemical reaction between a fuel and air to generate power also promises great efficiency gains, and holds out the possibility of eventually reducing carbon emissions completely if it ever becomes practicable to use a fuel such as hydrogen on a large scale (Breslin and Wang 2004, p. 36). Even using current NATO-standard F-76 marine diesel fuel, it is estimated that the use of fuel cells could reduce carbon dioxide emissions by 30 per cent and emissions of nitrogen oxides, carbon monoxide and hydrocarbons by 96 per cent (Office of the Under Secretary of Defense for Acquisition and Technology 2001, pp. 56–57). An ONR Fuel Cell Program has been under way for two decades, developing fuel cell technology for a number of different applications, including shipboard electrical power generation and unmanned vehicles (US Office of Naval Research 2008). The ONR's Naval Research Laboratory has already developed and successfully flown a small hydrogen fuel cell-powered UAV, named the Ion Tiger. Test flights have exceeded 24 hours in duration, carrying a 6 lb payload and producing no emissions (other than water vapour) at 'a low cost' and with low noise and heat signatures (US Office of Naval Research 2009b).

Research is also under way in the US on new aircraft turbine engine technologies, with the objective of improving fuel efficiency compared to the newest engines types in service by 25 per cent (Office of the Under Secretary of Defense for Acquisition, Technology and Logistics 2008, p. 42). In addition, new water jet technology may improve efficiency by up to 50 per cent over current models, with the improved hydrodynamic qualities of contra-rotating propellers able to improve ship propulsion efficiency by five to ten per cent. Further, the use of electro-mechanical actuators may lead to significant weight and personnel savings, thereby reducing fuel use on some classes of ship (Office of the Under Secretary of Defense for Acquisition, Technology and Logistics 2008, pp. 47–49). Future classes of ships are also likely to employ advances in hull form to reduce hydrodynamic drag and thus fuel consumption; fuel efficiency gains in the 5–10 per cent range are expected. Fuel consumption is directly affected by vessel weight, and the integration of advanced materials into future ships may reduce hull weights in the order of 30–50 per cent; topside structures could be reduced in weight by 25 to 35 per cent (Office of the Under Secretary of Defense for Acquisition and Technology 2001, p. 57).

The 2008 Defense Science Board report also identified three potential 'breakthrough' technologies among hundreds evaluated for their possible contribution to fuel efficiency. Of these, one is particularly relevant to maritime forces: the variable speed tilt rotor aircraft. Current studies indicate that advanced tilt rotorcraft could achieve at least twice the aerodynamic cruise efficiency of the current US V-22 tilt rotor and up to four times that of current helicopters. This translates to a 50 per cent reduction in fuel usage for long-distance or persistent operations, such as ISR missions conducted for MDA (Office of the Under Secretary of Defense for Acquisition, Technology and Logistics 2008, pp. 39–40). Further, as noted above, the use of unmanned vehicles will continue to expand rapidly, including in the maritime environment, with the types of missions envisioned for robotic craft also likely to expand. For example, while the ISR mission is currently the focus of most maritime UAV development, experimental unmanned combat aerial vehicles are already in an advanced stage of development in the US, including one prototype, the Northrop Grumman X-47B, intended for experimental operations from aircraft carriers (Scott 2010, p. 6; Singer 2009, p. 6). New technologies are also enabling modern interpretations on an old theme: the lighter-than-air vehicle (LAV), or airship, seems to be making a comeback, both in manned and unmanned modes. In the maritime environment, LAVs have great potential as low cost, highly efficient platforms, both for the ISR role and for transportation (Streetly 2010, pp. 52–55).

In terms of new types of anti-fouling coatings being developed by or for navies, the non-biocide silicone-based Intersleek 700 and newer fluoropolymer Intersleek 900 technologies marketed by paint manufacturer, International Paint, provide a representation of the products now available. The Royal Navy has applied Intersleek 700 to four frigates and three Type 45 destroyers since 2005, while Intersleek 900 has more recently been applied to three Type 45 destroyers, the aircraft carrier *HMS Ark Royal* and 10 other warships. The Spanish Navy has also progressively been coating its warships with Intersleek 900 since 2007. The company claims impressive resultant fuel savings, and emissions reductions of around 9 per cent, as well as an increase in top speed by 2–3 knots depending on the individual ship. These claims appear to be supported by Royal Navy operational experience (International Paint 2009a, 2009b; UK Ministry of Defence 2009). In the US, among the ONR-funded non-toxic advanced technology anti-bio-fouling solutions being developed is a patterned coating, trademarked as 'Sharklet', which 'mimics the inherent texture and antimicrobial properties of shark skin'. Another is based on zwitterionic (mixed-charge) compounds, which 'manipulate surface environments at the molecular level to prevent proteins from binding to the ship's surface' (US Office of Naval Research 2009a).

Lastly, in addition to all the fuel efficiency methods and technologies discussed above, it should be noted that shipboard refrigeration and cooling currently relies on two powerful greenhouse gases: chlorofluorocarbons (CFCs) and hydrofluorocarbons (HFCs). In the case of the US, as argued by Breslin and Wang (2004), although the quantity of these gases emitted by the US Navy is extremely small as a proportion of the service's total emissions of greenhouse gases, both have significant environmental drawbacks. Indeed any production of CFCs was banned internationally under the 1987 Montreal Convention due to their ozone-depleting characteristics. The US Navy does not expect to be CFC-free until 2050, and the replacement HFCs, while only released into the atmosphere in very small amounts, have a 'global warming potential' as much as 1300 times that of the predominant greenhouse gas, carbon dioxide (Breslin and Wang 2004, p. 36). As a result, and perhaps anticipating possible future international moves to curtail HFC use, new technologies are being pursued in the US in the areas of thermoelectric and thermoacoustic cooling and magnetic refrigeration (Breslin and Wang 2004, pp. 36–37).

The cumulative gains in fuel efficiency and other measures to reduce greenhouse gas emissions both in current and, especially, in future generations of ships and aircraft have thus been demonstrated to be substantial.

4. THE FUTURE FOR MARITIME SECURITY FORCES: CONTINUITY OR CHANGE?

What then, is likely to be the net impact of climate change factors for the planning considerations of maritime security forces? Seven interrelated arguments are tendered below to conclude this chapter.

First, the specific, often localized impacts of climate change are both uncertain in detail and likely to be incremental, rather than abrupt in effect. Consequently, navies and other maritime security forces should have sufficient time in which to consider their respective needs as the security-related impacts of global warming become more apparent. However, since planning for future force structure tends to be a protracted process, especially for larger, more sophisticated or expensive platforms, it is sensible to incorporate climate change considerations into strategic and force planning now.

Second, climate change will only be one factor among many that will shape future security environments, and navies generally will want to build and maintain the most capable forces their respective governments are willing to provide, as a hedge against the constant threat of a worsening strategic situation and other contingencies. As this essay has argued, climate change has the potential to, at least indirectly, generate demand for all three naval roles (diplomatic, constabulary and military), applicable across the entire spectrum of threat intensity. Correspondingly, the *Australian Maritime Doctrine* justifiably restates the naval force planning truism: 'The ability of warships to undertake the constabulary and diplomatic roles depends substantially on their ability to carry out their combat roles. The capability to do all these things is thus largely a by-product of the resources and core skills developed for warfighting' (Royal Australian Navy 2010, p. 99).

It would be unwise for navies to attempt to optimize their force structures solely, or even primarily, for expected security implications resulting from a warming planet, as predicting the future security environment with any degree of specificity or certainty is impossible. In this situation of uncertainty, particularly in terms of the downstream security implications of the physical warming effects, or even of these effects themselves, upon the natural environment, prudence rather than panic ought be the operative principle for policymakers and defence planners. In this case, prudence requires assessment, as far as it is practicable or possible, of potential security-related consequences of different plausible climate change outcomes over medium- and long-term periods, both within territories and waters under national jurisdiction, and in external regions of particular national interest or concern. However, such considerations also need to include planning for worst case, abrupt climate change scenarios, just in case the improbable occurs.

Navies fortunate enough to have been able to develop a 'balanced', or at least a reasonably balanced, force structure – that is, one that is flexible, 'with the range of capabilities required to . . . meet national security and military strategic goals' and 'to carry out a particular mission without unnecessary redundancy' (Royal Australian Navy 2010, p. 184) – would benefit from the maintenance of such balance into the future as a prudent hedge against the operation of contingency across the spectrum of maritime tasks, including those related to climate change. Where appropriate, other more specialized navies may also find it prudent to move to a more balanced force structure; and similarly, larger coast guard organizations may want to ensure sufficient balance within the context of their more narrow range of roles and responsibilities. To maximize national effectiveness, joint and inter-agency cooperation is preferable. This can be narrow, in terms of hulls in the water, such as exemplified by the US Navy–Coast Guard 'national fleet' concept (Allen and Mullen 2006, pp. 16–20; Department of the Navy and United States Coast Guard 2006). Conversely, it can be broad, in terms of whole-of-government approaches to MDA (Border Protection Command 2008, p. 6; Rahman 2010, pp. 214–219). Beyond national borders, combined (that is, multinational) operations are also becoming increasingly important and can increase the capabilities brought to bear on particular problems or crises, such as major natural disasters. As previously argued, interoperability is a key factor for the effectiveness of multinational operations: planning and exercising will be required to maximize the potential of such combined forces.

Third, many elements of maritime security force structure are inherently flexible in employment; for example, amphibious ships and aircraft carriers, surface combatants, OPVs and other larger patrol vessels, such as coast guard cutters. Therefore, ensuring that improved flexibility is built into future force structure without compromising core capabilities is recommended for force capability planners of both navies and the larger and more capable coast guard organizations, such as those of the US, Japan and India. As a rule, maximizing ship flexibility means increasing its size. It may well be that for many coastal States, climate change considerations may eventually require an enlarged, or at least a more effective, force for the constabulary function, including for both the MDA and enforcement tasks. There will thus be implications for States that operate coast guards or similar non-military forces for such purposes. However, it is worth remembering that even the most capable coast guards do not have the types of vessels or capabilities necessary for many constabulary and related diplomatic roles. Navies will likely continue to be monopoly providers of particular maritime capabilities, such as for those instances in which the most sophisticated ISR assets, amphibious ships and medium and heavy-lift rotorcraft are required.

Fourth, while the arguments made above apply broadly, each individual State must tailor its future maritime security force structure to its specific needs. In many developing States, there already exists a patrol asset deficit. Further, it is likely that negative climate change effects will only increase the level of illegal activity at sea throughout, for example, archipelagic Southeast Asia. While patrol boats can be procured relatively quickly and cheaply, larger assets take more time to finance, construct and to become operationally effective, and there is the additional need to train crews and maintenance workers. Another asset that would be of benefit in States such as Indonesia and the Philippines is the amphibious ship. However, in such cases the aforementioned rule of 'bigger is better' may not necessarily apply. There may be considerable value in procuring greater numbers of smaller ships, rather than fewer larger ones, due to the nature of the geography of these regions, which boast many thousands of islands spread over large areas, with the corresponding need for capabilities to be deployed throughout the archipelago. The limited capacity of such States to afford or operate large ships is another reason that smaller ships would be more suitable in these cases.

In contrast, in India, large amphibious ships and aircraft carriers will likely be very important. This is because, in this part of the world, large numbers of people live in low-lying coastal areas and are highly vulnerable to sea level rise and, more urgently, storm surges and related weather events (Intergovernmental Panel on Climate Change 2007, p. 484). This is true for not only parts of India, but also for other parts of South Asia, such as Bangladesh and the Maldives. Owing to India's expanding role as a regional power, it can be expected to assume further responsibilities as a first responder to crises throughout much of the Indian Ocean region.

Fifth, comprehensive MDA is growing in value for all coastal States; it is not just ISR or vessel-related data that will be important. Maritime security forces are both generators and consumers of oceanographic, hydrographic and meteorological information. This demand for information on the physical characteristics of the maritime environment will only increase in response to changes to that environment caused by global warming. Indeed, as already noted, the *US Navy Climate Change Roadmap* has observed that improved environmental assessment in a changing climate is vital for tactical purposes through to long-term policy planning. The implication of this point is that further investment will be needed both in whole-of-nation marine scientific and climate research, and in hydrographic and oceanographic ships, whether they are operated by navies or civilian agencies.

Sixth, mitigation of the (relatively limited) climate-change impact generated by navies and other maritime security forces is only in its infancy. However, as argued in this chapter, the dual imperative to reduce fuel costs and improve energy security has the advantage of reducing greenhouse gas emissions. Although much more can be accomplished to improve the efficiency of existing vessels, the more significant gains will be made over the medium term as new technologies are integrated into future ship and aircraft designs. The US Navy, not for the first time, is the technology leader in this regard, with the potential to instigate real advances in emissions reductions, applicable not just to navies, but also to the civilian economy. In the Asia-Pacific region, these efforts will be uneven; it is not clear whether developing States will share the same sense of responsibility as exists in many of the advanced economies to require their maritime security forces to address the issue of climate change. Perhaps high fuel prices, which tend to negatively affect the operational capacity of developing State maritime forces to an even greater extent than rich ones, will play a role here. Further reducing the likelihood that developing nations will be capable of participating in emissions reduction, some of the new technologies will initially be expensive, potentially outweighing the benefit of the long-term savings they generate. Yet others may be too sensitive to sell to certain politically unstable rivals.

Finally, the overall impact of climate change on maritime security forces is likely to be relatively easy to manage and, as with climate change itself, slow and incremental rather than abrupt. There will be more continuity than change for policy and force planners, although planning will be essential for the longer term. Maritime forces will eventually need to become 'greener' and understand the potential demand for their services generated by the effects of climate change. If anything should be clear for policymakers and maritime force planners from the analysis presented in this chapter, it is that there is a need for further investment in knowledge and understanding. This investment should encompass a variety of areas, including basic scientific research and marine environmental assessment, the development of new technological solutions to emissions reductions, and better national and regional security risk assessments. This is the only prudent course of action.

REFERENCES

ABC Radio Australia (2010), 'Oil spill victims turn to people smuggling', ABC Radio Australia transcript, 12 August.

Allen, Admiral Thad, USCG and Admiral Mike Mullen, USN (2006), 'America's national fleet: a coast guard–navy imperative', US Naval Institute *Proceedings*, August.

Bailey, J.J. (2009), 'Is it practical for defence to reduce its carbon emissions without affecting its effectiveness?' *Defence Studies*, **9** (1), 69.

Barnett, Jon and W. Neil Adger (2007), 'Climate change, human security and violent conflict', *Political Geography*, **26** (6), 639–655.

Bateman, Sam (2003), 'Coast guards: new forces for regional order and security', *AsiaPacific Issues*, **65**.

Biello, David (2009), 'Navy green: military investigates biofuels to power its ships and planes', *Scientific American*, 14 September.

Billings, Kevin, David R. Oliver and George Alan Hutchinson (2008), 'Get on board with alternative fuels', US Naval Institute *Proceedings*, December.

Booth, Ken [1977] (1979), *Navies and Foreign Policy*, New York: Holmes & Meier.

Border Protection Command (2008), *Concept of Operations*, Canberra: BPC.

Breslin, David A. and Ye-Ling Wang (2004), 'Climate change, national security, and naval ship design', *Naval Engineers Journal*, **116** (1), 29–30.

CNA Corporation (2007), *National Security and the Threat of Climate Change*, Alexandria, VA: CNA Corporation.

Department of the Navy and United States Coast Guard (2006), 'National fleet: a joint navy/coast guard policy statement', 3 March.

Dupont, Alan and Graeme Pearman (2006), *Heating Up the Planet: Climate Change and Security*, Lowy Institute Paper 12, Sydney: Lowy Institute for International Policy.

Elleman, Bruce A. (2007), *Waves of Hope: The U.S. Navy's Response to the Tsunami in Northern Indonesia*, Newport Paper No. 28, Newport, RI: Naval War College Press.

English, Colin R. (2003), 'The WR-21 intercooled recuperated gas turbine engine – integration into future warships', *Proceedings of the International Gas Turbine Congress 2003*, 2–7 November, Tokyo.

Fish, Tim (2010), 'Striking distance', *Jane's Defence Weekly*, 16 June, pp. 22–27.

Fish, Tim, Edward Lucas and Jon Rosamond (2008), 'Arctic countries vying for a piece of the ice seek to strengthen regional presence', *Jane's Navy International*, September, pp. 12–14

Funnell, Cliff (2009), 'Safe and sound: UUV-based sonar gets closer to mines', *Jane's Navy International*, June, pp. 26–31.

German Advisory Council on Climate Change (2008), *Climate Change as a Security Risk*, London: Earthscan.

Grove, Eric (2007), 'Carrier waves: programmes speak of an enduring appeal', *Jane's Navy International*, November, pp. 31–32.

HM Government (2010), *Securing Britain in an Age of Uncertainty: The Strategic Defence and Security Review*, London: The Stationery Office.

Hobson, Sharon and Casandra Newell (2009), 'Shrinking ice cover creates opportunities and threats', *Jane's Navy International*, January/February, pp. 16–21.

Hsu, Tiffany (2010), 'Navy gets 1,500 gallons of algae-based fuel from solazyme', *Los Angeles Times*, 19 July.

Intergovernmental Panel on Climate Change (2007), *Climate Change 2007: Impacts, Adaptation and Vulnerability*, Contribution of Working Group II to the Fourth Assessment Report of the Intergovernmental Panel on Climate Change, Cambridge: Cambridge University Press.

International Maritime Organization [1973] (1978), *International Convention for*

the Prevention of Pollution from Ships, 1973, as modified by the Protocol of 1978 (MARPOL 73/78).
International Maritime Organization [1973, 1978] (1997), Regulations for the Prevention of Air Pollution from Ships – Protocol to MARPOL 73/78 (Annex VI).
International Maritime Organization [1973, 1978] (2008), Revised MARPOL Annex VI, 10 October.
International Paint (2009a), 'The Spanish Navy opts for Intersleek® 900', accessed 26th July 2012 at: www.international-marine.com.
International Paint (2009b), 'UK Navy flagship gets coated with Intersleek® 900', 7 September, accessed 26th July 2012 at: www.international-marine.com.
Kerr, Glenn and Barry Snushall (eds) (2005), *Future Environmental Policy Trends to 2020: Impact on Ship Design and Operations*, Papers in Australian Maritime Affairs No. 13, Canberra: Sea Power Centre – Australia.
Levy, Marc A. (1995), 'Is the environment a national security issue?' *International Security*, **20** (2), 58.
Mabus, Ray, The Honorable Secretary of the Navy (2009), Remarks to the Naval Energy Forum, McLean, VA, 14 October.
Mak, J.N. (2007), 'Pirates, renegades, and fishermen: the politics of "sustainable" piracy in the strait of Malacca', in Peter Lehr (ed.), *Violence at Sea: Piracy in the Age of Global Terrorism*, New York: Routledge, p. 200.
Markle Stephen P. and Sean E. Gill (2006), 'T-AKE: acquiring the environmentally sound ship of the 21st century', *Naval Engineers Journal*, **118** (3), pp. 89–91.
Minnick, Wendell (2009), 'US forces on Taiwan for typhoon relief mission', *Defense News*, 19 August.
Office of the Under Secretary of Defense for Acquisition and Technology, Report of the Defense Science Board (2001), *More Capable Warfighting through Reduced Fuel Burden*, Washington, DC.
Office of the Under Secretary of Defense for Acquisition, Technology and Logistics (2008), *Report of the Defense Science Board Task Force on DoD Energy Strategy: 'More Fight – Less Fuel'*, Washington, DC.
Ohls, Gary J. (2009), *Somalia . . . From the Sea*, Newport Paper No. 34, Newport, RI: Naval War College Press.
Pengelley, Rupert (2009), 'All hands on deck: the sky's the limit for shipboard UAVs', *Jane's Navy International*, June, pp. 12–17.
Pumphrey, Carolyn (ed.) (2008), *Global Climate Change: National Security Implications*, Carlisle, PA: Strategic Studies Institute, U.S. Army War College.
Rahman, Chris (2008), *The Global Maritime Partnership Initiative: Implications for the Royal Australian Navy*, Papers in Australian Maritime Affairs No. 24, Canberra: Sea Power Centre –Australia.
Rahman, Chris (2009), *Concepts of Maritime Security: A Strategic Perspective on Alternative Visions for Good Order and Security at Sea, with Policy Implications for New Zealand*, Discussion Paper No. 07/09, Wellington, New Zealand: Victoria University of Wellington, Centre for Strategic Studies.
Rahman, Chris (2010), 'Maritime domain awareness in Australia and New Zealand', in Natalie Klein, Joanna Mossop and Donald R. Rothwell (eds), *Maritime Security: International Law and Policy Perspectives from Australia and New Zealand*, London: Routledge.
Reuters, 25 August 2010, 'US Navy Floats an Ambitious Energy Plan [Interview]'.
Rolls-Royce (n.d.) 'WR-21 Marine Gas Turbine', accessed 26th July 2012 at:

http://www.rolls-royce.com/marine/products/diesels_gas_turbines/gas_turbines/wr21.jsp.
Rosamond, Jon (2008), 'All systems go as electric solutions power future ships', *Jane's Navy International*, May, pp. 20–25.
Rosenthal, Elisabeth (2010), 'US military orders less dependence on fossil fuels', *The New York Times*, 4 October.
Royal Australian Navy (2010), *Australian Maritime Doctrine: RAN Doctrine 1, 2010*, 2nd edn, Canberra: Sea Power Centre – Australia.
Scott, Richard (2007), 'Power trip: energising naval electric weapons research', *Jane's Navy International*, March, pp. 34–39
Scott, Richard (2008), 'New-generation USVs offer increased mission diversity', *Jane's Navy International*, April, pp. 16–22.
Scott, Richard (2009), 'Light fantastic: US eyes laser weapon technology for future ship self-defense', *Jane's Navy International*, June, pp. 8–10.
Scott, Richard (2010), 'Technical glitches delay UCAS-D test', *Jane's Navy International*, January/February, p. 6.
Singer, P.W. (2009), *Wired for War: The Robotics Revolution and Conflict in the 21st Century*, New York: The Penguin Press.
Streetly, Martin (2010), 'Eyes down: at last a viable role for airships?' *Jane's International Defence Review*, November, pp. 52–55.
The President of the United States of America (2009), Executive Order 13514 – Federal Leadership in Environmental, Energy, and Economic Performance, 8 October.
Till, Geoffrey (2009), *Seapower: A Guide for the Twenty-First Century*, 2nd edn, London: Routledge.
Tiron, Roxana (2010), 'Fuel's Full Burden', *Seapower*, March.
UK Ministry of Defence (2009), 'A greener and leaner HMS Ark Royal returns to sea', MOD press release, 25 September.
UK Ministry of Defence (2010a), *MOD Climate Change Strategy 2010*, April.
UK Ministry of Defence (2010b), *Sustainable Procurement Strategy*.
United Nations (1982), *United Nations Convention on the Law of the Sea, 1982*, Article 236.
US Department of Defense (2008), *Report to Congress on Energy Security Initiatives*, DoD Energy Security Task Force.
US Department of Defense (2010), *Quadrennial Defense Review Report*, February.
US National Intelligence Council (2009), 'The Impact of Climate Change to 2030', accessed 26th July 2012 at: http://www.dni.gov/nic/special_climate2030.html.
US Naval Institute (2005), US Naval Institute *Proceedings*, pp. 10–19.
US Naval Sea Systems Command (2010), 'MAR04-03: PEO ships participates in ASNE energy futures symposium', *NAVSEA newswire*.
US Navy Chief of Naval Operations (2009), *US Navy Arctic Roadmap*, Task Force Climate Change/Oceanographer of the Navy.
US Navy Chief of Naval Operations (2010), *US Navy Climate Change Roadmap*, Task Force Climate Change/Oceanographer of the Navy.
US Office of Naval Research (2008), 'Fuel cell program', ONR information sheet.
US Office of Naval Research (2009a), 'Biofouling prevention coatings', ONR information sheet.
US Office of Naval Research (2009b), 'Ion Tiger', ONR information sheet.
US Office of Naval Research (2009c), 'New hull coatings for navy ships cut fuel use, protect environment', ONR press release, 4 June.

Walsh, Edward J. (2010), 'How green is my ship?', US Naval Institute *Proceedings*, February.
Weir, Gary E. (2010), 'Fish, family, and profit: piracy and the horn of Africa', in Bruce A. Elleman, Andrew Forbes and David Rosenberg (eds), *Piracy and Maritime Crime: Historical and Modern Case Studies*, Newport Paper No. 35, Newport, RI: Naval War College Press, pp. 207–221.
Williams, Meryl J. (2007), *Enmeshed: Australia and Southeast Asia's Fisheries*, Lowy Institute Paper 20, Sydney: Lowy Institute for International Policy.

9. Climate change and shipping

Stathis Palassis

1. INTRODUCTION

The reduction of greenhouse gas (GHG) emissions and the stabilization of global climate is one of the greatest international challenges.[1] The task is enormous as States transform their carbon-based economies into a low-carbon sustainable global economy. Despite the apparent difficulties involved, the longer action is delayed, the more difficult the challenge becomes. The greenhouse challenge captures all aspects of the energy, industry and transport sectors, including the reduction of GHG emissions from the international shipping sector. In 2007 international shipping is estimated to have contributed around 2.7 per cent of the total global carbon dioxide (CO2) emissions (IMO 2011a). International shipping underpins the global economy and is the most cost-effective and environmentally efficient of all forms of transport (IMO 2005). However, as significant growth in commercial shipping is expected in response to increased international trade, ship-source pollution is expected to rise dramatically. It thus becomes imperative that the international shipping sector take steps to minimize its GHG emissions.

The International Maritime Organization (IMO) has been actively considering the development of rules for the reduction of the GHG emissions of international shipping as part of the larger task of reducing ship-sourced air pollution. Despite having successfully regulated some ship-sourced air pollutants,[2] reaching consensus within the Organization on the regulation of the sector's GHG emissions has been difficult. The IMO is promoting a global approach to all its initiatives, including the reduction of GHG emissions, while developing States have been seeking differentiated responsibility over the reduction of their GHG emissions.[3] The conflicting positions of developed and developing States have meant that reduction proposals have encountered difficulties in being adopted. Irrespective of these differences, significant progress has been made in adopting the regulations necessary for the reduction of GHG emissions from the international shipping sector, namely the creation of technical and operational measures and suitable market-based measures (MBMs).

Topics that will be covered in this chapter include: the international legal framework to reduce GHG emissions; the international legal framework for the prevention, reduction and control of ship-sourced marine pollution; the international regulation of ship-sourced air pollution of the marine environment; and the international regulation of ship-sourced GHG emissions. In the section dealing with the regulation of ship-sourced GHG emissions, discussion will focus on the three pillars of the IMO's work; namely, technical and operational measures and market-based measures (MBMs).

2. THE INTERNATIONAL LEGAL FRAMEWORK TO REDUCE GREENHOUSE GAS EMISSIONS

The international legal framework for the reduction of GHG emissions operates within the various multilateral agreements adopted through the United Nations (UN) law-making process, which have facilitated the development of specialist climate change forums and processes. Despite this, treaty making for the reduction of GHG emissions and the greater climate change challenge has been a difficult and complex process, as the rules that need to be created strike at the heart of the notion of economic prosperity and require States to adopt a significant shift in their energy, industrial and transport law and policies.

2.1 The United Nations Framework Convention on Climate Change

The United Nations Framework Convention on Climate Change (UNFCCC 2011a) was adopted in 1992 at the United Nations Headquarters in New York and was then opened for signature and ratification in June of the same year at Rio de Janeiro within the auspices of the United Nations Conference on Environment and Development (UNCED). Even though the UNFCCC is a framework, and aspirational in character, the treaty proved difficult to negotiate, as States were generally reluctant to create legally binding obligations to reduce their GHG emissions and to be bound by principles of sustainable development, particularly in relation to the precautionary principle.[4] The challenges experienced were primarily related to, firstly, scientific uncertainty over what some perceived to be unsettled climate change science; secondly, the associated economic costs of adopting GHG reduction measures; and thirdly, the distribution of responsibility over reduction measures between developed and developing

States. Despite these difficulties, the UNFCCC entered into force on 21 March 1994 and 165 States have now ratified the treaty (UNFCCC 2011c). The objective of the Convention is stated as:

> to achieve, in accordance with the relevant provisions of the Convention, stabilization of greenhouse gas concentrations in the atmosphere at a level that would prevent dangerous anthropogenic interference with the climate system. Such a level should be achieved within a timeframe sufficient to allow ecosystems to adapt naturally to climate change, to ensure that food production is not threatened and to enable economic development in a sustainable manner. (UNFCCC 2011c, art 2)

The UNFCCC establishes, among other things: commitments to stabilize GHG concentrations in the atmosphere to a safe level; a financial mechanism and a commitment by developed States to provide financial resources towards this objective; two subsidiary bodies, the Conference of the Parties (COP) and the Atmospheric Protection Agency (APA); a number of guiding principles; and innovative implementation and dispute settlement mechanisms.

Of particular significance is the position of developing States and their responsibilities under the UNFCCC. The treaty adopted provisions on the principle of common but differentiated responsibilities (CBDR) in the reduction of GHG emissions, which are included both in the treaty's preamble and in the substantive text:

> Acknowledging that the global nature of climate change calls for the widest possible cooperation by all countries and their participation in an appropriate and effective international response, in accordance with their common but differentiated responsibilities and respective capabilities and their social and economic conditions. (UNFCCC 2011a, Preamble)

> The Parties should protect the climate system for the benefit of present and future generations of humankind, on the basis of equity and in accordance with their common but differentiated responsibilities and respective capabilities. Accordingly the developed country Parties should take the lead in combating climate change and the adverse effects thereof. (UNFCCC 2011a, art 3(1))

Despite the UNFCCC's absence of binding reduction targets for GHG emissions, it does contain a series of commitments commencing with a further recognition of CBDR by the treaty, which is worded as: 'All parties, taking into account their common but differentiated responsibilities and their specific national and regional development priorities, objectives and circumstances, shall . . .' (UNFCCC 2011a, art 4). The treaty contains strategic and procedural commitments (UNFCCC 2011a, art 4(1)), while at the same time creating obligations specifically for industrialized States, contained in Annex I of the treaty.[5] Annex I States are required to adopt national policies and to take measures on the mitigation of climate change by limiting their anthropogenic emissions of GHGs and by protecting and enhancing their GHG sinks and reservoirs. Therefore, there is a clear duty on the part of Annex I States to mitigate climate change. However, discretion is given to these States as to which mechanisms will be utilized to restrict GHG emissions and enhance GHG sinks and reservoirs.

The UNFCCC advocates the adoption of further international agreements to reduce GHG emissions and it was thus the all-important first step in the international legal response to the climate change challenge.

2.2 The Kyoto Protocol

The Kyoto Protocol to the UNFCCC (Kyoto Protocol) was adopted on 11 December 1997 at Kyoto during the COP 15 of the UNFCCC (UNFCCC 2011b). Despite the absolute necessity of facilitating universal cooperation to combat climate change through the reduction of GHG emissions, the Kyoto Protocol was an extremely difficult agreement to negotiate. The major obstacle was the role of developing States and whether they should be included alongside developed States in being legally bound to reduce their GHG emissions. The exclusion of developing States from the first set of binding reduction targets led to the non-ratification of the Protocol by several developed States, including the United States and Australia.[6] The role of developing States in accepting obligations to reduce their GHG emissions alongside developed States thus became the most difficult aspect of applying the principle of CBDR. However, it is important to note that the GHG emissions of developing States are rising dramatically and it is projected that during the next decade their net GHG emissions will exceed those of developed States. In addition, the Kyoto Protocol's ratification requirements were onerous in that it required a significant number of ratifications from developed States to commence operation; that is, the Protocol required ratification by 55 States and sufficient industrialized states to account for 55 per cent of global carbon emissions according to a 1990 baseline (Kyoto Protocol, art 25). As a consequence, the Kyoto Protocol did not commence operation until 16 February 2005. A total of 84 States have now ratified the Protocol (UNFCCC 2011c).

The Kyoto Protocol creates binding reduction targets for the emission of GHGs by means of two annexes: Annex 'A' and Annex 'B'. Annex 'A' provides for a group of six GHGs that are to be the subject of binding reduction targets. These are: CO2, methane, nitrous oxide, hydrofluorocarbons, perfluorocarbons and sulphur hexafluoride. Annex 'A' also sets out relevant sectors, including the transport sector, as well as source categories of the air pollutants. Annex 'B' lists the reduction targets for 38 States, including the industrialized States and the European Community. The Protocol operates by placing a cap on the aggregate amount of GHGs that each industrialized State can emit. Article 3 of the Protocol requires industrialized States to reach an overall reduction of GHG emissions to 5 per cent below 1990 levels for the period 2008–2012, which is to be effected through a variety of mechanisms. Each developed State must ensure that their aggregate emissions do not exceed the assigned amount (Kyoto Protocol, art 3).[7] The year 1990 is used as the baseline from which current carbon emissions are measured, and each developed State is given a different percentage by which they must adjust their GHG emissions relative to their 1990 levels. Most States have to reduce GHG emissions, while others, such as Australia, were able to negotiate an increase. These adjustment levels are intended to ensure that the overall GHG emissions from Annex 'B' States do not exceed the set target of at least five per cent below 1990 figures for the period 2008– 2012. Emissions are measured as an average over five years from 2008 to 2012, primarily because of an amendment owing to United States' concerns over aberrant weather conditions occurring within a single year that could lead to an abnormal result.

The exclusion of developing States from Annex 'B' and the use of differential targets were and continue to be highly contentious. In addition to the Kyoto Protocol containing the obligation to achieve quantified emission reductions based on 1990 levels over the period 2008–2012, the Protocol provides three other mechanisms for assisting States to meet their reduction targets. These are market-based mechanisms comprising: an emissions trading scheme (ETS); a clean development mechanism (CDM); and joint implementation (JI). 'The mechanisms help stimulate green investment and help Parties meet their emission targets in a cost-effective way' (UNFCCC 2011d).

As mentioned, articles 2 and 10 of the Kyoto Protocol include obligations for the transport sector. Article 2(2) directs States to operate through the IMO for the reduction of ship-sourced GHG emissions, while Article 10 reaffirms that these rules need to reflect the principle of CBDR:

> The Parties included in Annex I shall pursue limitation or reduction of emissions of greenhouse gases not controlled by the Montreal Protocol from aviation and marine bunker fuels, working through ICAO and IMO, respectively (Kyoto Protocol art 2(2)).

> All Parties, taking into account their common but differentiated responsibilities and their specific national and regional development priorities, objectives

> and circumstances, without introducing any new commitments for Parties not included in Annex I, but reaffirming existing commitments under Article 4 paragraph 1, of the Convention, and continuing to advance the implementation of these commitments in order to achieve sustainable development, taking into account Article 4, paragraphs 3, 5 and 7 of the Convention, shall: . . . (Kyoto Protocol, art 10)

The Parties to the Kyoto Protocol at the COP 12 at Nairobi, Kenya adopted, on 17 November 2006, an Amendment to Annex 'B' of the Kyoto Protocol to the UNFCCC (UNFCCC 2006, Decision 10/CMP 2). However, this amendment is not significant to either international shipping or to the transport sector more generally.

2.3 The Precautionary Principle

At the 1992 UNCED, States adopted the Rio Declaration on Environment and Development (Rio Declaration). Principle 15 of the Rio Declaration requires States to take a precautionary approach:

> In order to protect the environment, the precautionary approach shall be widely applied by States according to their capabilities. Where there are threats of serious or irreversible environmental damage, lack of full scientific certainty should not be used as a reason for postponing cost-effective measures to prevent environmental degradation.

The precautionary approach 'ensures that substances or activities posing environmental threat are prevented from adversely affecting the environment even if no conclusive scientific proof links the particular substance or activity to the environmental damage' (Cameron and Abouchar 1991, p. 2). It requires States to undertake environmental protection action, even though full scientific knowledge is lacking as to the effects of an activity, particularly considering that the repair of environmental damage costs more than its prevention. The precautionary approach encourages decision-makers to consider all the potential environmental impacts of development and requires that decision-makers err on the side of caution to ensure no negative environmental impacts can occur. It remains unsettled as to whether the precautionary approach is an overarching principle of policy and decision-making, or whether it is an approach that is utilized only for hazardous and ultra-hazardous activities. The distinction may explain the variations in precaution adopted by global and regional treaties. This distinction has prohibited the definition of the precautionary approach as provided by Principle 15 of the Rio Declaration from emerging as a rule of customary international law (Palassis 2011 pp. 64–65).

2.4 Post-Kyoto Protocol Initiatives

In 2007, the States Parties to the UNFCCC agreed at the Bali Conference to adopt the Bali Action Plan to meet the immediate challenges of climate change; namely, adaptation, mitigation, technology transfer and financing. They agreed to shape an ambitious and effective international response to climate change at the Copenhagen Conference. Regrettably, the Copenhagen Conference of 2009 did not adopt another treaty to replace the Kyoto Protocol. However, the Conference did adopt the Copenhagen Accord, which emphasizes that combating climate change is one of the greatest challenges of the current period and that action consistent with the principle of CBDR and respective capabilities is urgently needed (Copenhagen Accord, art 1).

The last meeting of the Parties to the UNFCCC was the COP 16, which was held at Cancun in Mexico in December 2010. At that meeting, there was discussion on the provision of predictable and sustainable climate finance sources by 2013, worth at least US$100 billion a year, to assist the world's poorest people to cope with climate change and to rebuild solidarity between developed and developing States. Despite the Cancun Conference not making any specific decisions regarding climate change and the transport sector, it was agreed that the IMO should continue pursuing the reduction or limitation of GHG emissions from international shipping and that it should inform the UNFCCC Parties of the progress made since the Copenhagen meeting on all three pillars of the IMO's work; namely, the future adoption of technical, operational and market measures. Most importantly, the conclusions reached at the Cancun meeting affirmed that the IMO is the appropriate international forum within which to promulgate the rules for the reduction of GHG emissions from international shipping. The next meeting of the COP to the UNFCCC will be held at Durban in South Africa in December 2011.

3. THE INTERNATIONAL LEGAL FRAMEWORK FOR THE PREVENTION, REDUCTION AND CONTROL OF SHIP-SOURCED MARINE POLLUTION

The international legal framework for the prevention, reduction and control of ship-source marine pollution operates multilaterally within the framework provided by the international law of the sea.

3.1 The Law of the Sea Convention

Under the 1982 United Nations Convention on the Law of the Sea (LOSC), which came into force on 16 November 1994, States are obligated to reduce pollution of the marine environment from all sources. Part XII of the LOSC is devoted to the protection and preservation of the marine environment and recognizes that 'States have the obligation to protect and preserve the marine environment' (LOSC 1982, art 192). It also emphasizes the importance of uniform international standards in regulating marine pollution, with Article 194 providing that:

1. States shall take, individually or jointly as appropriate, all measures consistent with this Convention, that are necessary to prevent, reduce and control pollution of the marine environment from any source, using for this purpose the best practicable means at their disposal and in accordance with their capabilities, and they shall endeavor to harmonize their policies in this connection.
2. States shall take all measures necessary to ensure that activities under their jurisdiction or control are so conducted as not to cause damage by pollution to other states and their environment, and that pollution arising from incidents or activities under their jurisdiction or control does not spread beyond the areas where they exercise sovereign rights in accordance with this Convention.

Part XII is a framework, and sets out the competence of States over six sources of marine pollution: land-based (LOSC 1982, art 207); seabed activities subject to national jurisdiction (LOSC 1982, art 208); seabed beyond the limits of national jurisdiction (LOSC 1982, art 209); dumping (LOSC 1982, art 210); ships (LOSC 1982, art 211); and from or through the atmosphere (LOSC 1982, art 212). The LOSC defers details for the technical rules that are to be formulated, and directs that 'generally accepted international rules and standards' are to be preferred to national laws and regulations (Franckx 1997, p. 68).

3.2 The International Maritime Organization

The LOSC directs States to act through the competent international organization, or general diplomatic conference, in establishing international rules and standards to prevent, reduce and control pollution of the marine environment from ships (LOSC 1982, arts 207(4), 208(5), 210(4), 211(3) and 212(3)). The IMO is viewed as the competent international organization for ship-source pollution. Even though not explicitly referred to by the LOSC, the conduct of States indicates that the IMO is accepted as the forum within which these international rules and standards are to be developed. Further, as the IMO has been acting as the competent international organization, it demonstrates an acceptance of its functions by States.

The IMO is an autonomous and specialized UN agency, established by treaty in Geneva in 1948 to create international rules and standards for safety of navigation and pollution prevention.[8] Its activities are reflected by its mission statement of 'safe, secure and efficient shipping on clean seas'. The Marine Environment Protection Committee (MEPC) is a subsidiary body of the IMO Council comprised of all member States; it meets every nine months. It is empowered to consider any matter within the scope of the IMO's work concerned with the prevention, reduction and control of pollution from ships. Since its creation, the IMO has facilitated the successful international regulation of ship-source marine pollution. Particularly impressive has been the dramatic reduction of the quantity of oil spilled by ships (ICS) and the creation of liability and compensation regimes for ship-sourced marine pollution damage. So widely supported is the civil liability regime that it has been used as a model in the creation of other international liability regimes. The ship-source civil liability contains two additional regimes modelled on the oil spill civil liability regime, one in respect of bunker fuels, and one for hazardous noxious substances (IMO 2009a). Despite these regulatory successes, issues surrounding ship-source air pollution, including reduction of GHG emissions have, until recently, remained largely unregulated.

3.3 International Environmental Soft Law

The 1992 Rio Declaration does not contain any provisions specific to the protection and sustainable use of the oceans and seas. However, Principle 13 of the Rio Declaration provides that:

> States shall develop national law regarding liability and compensation for the victims of pollution and other environmental damage. States shall also cooperate in an expeditious and more determined manner to develop further international law regarding liability and compensation for adverse effects of environmental damage caused by activities within their jurisdiction or control to areas beyond their jurisdiction.

Chapter 17 of Agenda 21 is devoted to protection of the oceans and seas and sets out the basis for action for marine environmental protection (Agenda 21, paras 17.20–23). This is the most important action plan in respect of the oceans and seas. It provides that new approaches to marine and coastal area management and development need to be adopted[9] and deals with more specific matters including programmes to prevent, reduce and control pollution of the marine environment from sea-based sources (Agenda 21, paras 17.30–34). States are to act through the IMO and other relevant international organizations in assessing the need for additional measures to address degradation of the marine environment. In relation to the international shipping sector, States should be '[s]upporting the ongoing activity within IMO regarding development of appropriate measures for reducing air pollution from ships' (Agenda 21, para 17.30(a)(ix)).

4. THE INTERNATIONAL REGULATION OF SHIP-SOURCED AIR POLLUTION OF THE MARINE ENVIRONMENT

Air pollution from ships, in particular marine diesel engines, is a significant source of total global air pollution emitted. It is therefore also a major source of marine pollution. Further, increased traffic from ships in ports has meant significant increases in the on-shore air pollution contributed by the international shipping sector. As marine diesel engines burn the lowest grade diesel fuel and heavy oil, their emissions generate harmful exhaust fumes that significantly contribute to environmental degradation and harm public health (Bright Hub 2011). If no changes are made and ship-source air emissions continue to rise, it is anticipated that by 2020, air pollution from ships will exceed that from land-based sources. Ship-source air emissions have thus been the subject of much activity within the IMO.[10]

4.1 The International Convention for the Prevention of Pollution from Ships (MARPOL 73/78)

In response to a UN resolution requiring the IMO to promote the adoption of effective international environmental agreements on the prevention and control of marine pollution, the IMO in 1973 adopted the International Convention for the Prevention of Pollution by Oil from Ships (MARPOL 73). As the Convention's obligations and technical discharge criteria were too onerous, States did not readily ratify and thus the Convention could not enter into force. A further diplomatic conference in 1978 adopted the 1978 Protocol to MARPOL 73, the result of which was a substantially revised Convention known as MARPOL 73/78. The Convention is undoubtedly one of the most important multilateral treaties on the protection of the marine environment. The aim of MARPOL 73/78 is stated in its preamble as '[d]esiring to achieve the complete elimination of intentional pollution of the marine environment by oil and other harmful substances and the minimization of accidental discharges of such substances'. The

Convention applies to '[s]hips entitled to fly the flag of a party to the Convention and to non-flag State ships which operate under the authority of a party to the Convention' (MARPOL 73/78, art 3(1)). The Convention does not apply to any warship, naval auxiliary or other ship owned or operated by a State and used only on government non-commercial service (MARPOL 73/78, art 3(3)). The MARPOL 73/78 Convention is a comprehensive and highly technical treaty, combating six sources of ship-source marine pollution through its six regulatory annexes, which cover the following pollutants: oil, noxious liquid substances in bulk, harmful substances in packaged form, sewage, garbage and air pollution.

4.2 Annex VI of MARPOL 73/78: Air Pollution

Within the MARPOL 73/78 regime, air emissions have been the last pollution medium to be regulated. Despite the IMO commencing work in this area in the 1980s, its laws were not finalized until 26 September 1997 through the Protocol of 1997, which adopted Annex VI – Prevention of Air Pollution from Ships. Only States Parties to MARPOL 73/78 are able to become Annex VI Parties. Annex VI came into force on 19 May 2005, with 25 States Parties, representing 50 per cent of the world's registered tonnage. By February 2006, it had been ratified by 30 States, representing 63.73 per cent of the world's registered tonnage. A revised Annex VI was adopted on 10 October 2008 and came into force on 1 July 2010 (MEPC 2008, p. 58).

4.3 Application of Annex VI Measures

Annex VI of MARPOL 73/78 applies to all ships, drilling rigs and other platforms. Certification of Annex VI requirements is mandatory for all ships involved in international trade. Annex VI imposes strict limits on the emissions of oxides of nitrogen (NOx) and oxides of sulphur (SOx) from engines, while setting out accompanying rules on the demonstration of compliance. Further, fire fighting, refrigeration and incineration systems are all subject to regulatory limits. The limits and regulations contained in Annex VI were set at modest levels, probably to make the laws more acceptable to States. The laws apply to new engines only, so that manufacturers would have little problem in meeting the set limits. Emissions are any release of substances from ships that are subject to control under Annex VI into the atmosphere or sea.

Annex VI sets out four principal sets of regulatory requirements: firstly, it limits the emissions of NOx from marine diesel engines; secondly, it requires that ships avoid using fuel with a sulphur content exceeding 4.5

per cent by mass; thirdly, it prohibits the deliberate emissions of ozone-depleting substances (ODS); and fourthly, it prohibits the incineration of certain products on board ships. The various IMO air emission standards are generally referred to as Tier I, II and III standards. Tier I standards were defined through the 1997 adoption of Annex VI, while Tier II and III standards were introduced through Annex VI amendments adopted in 2008. In addition to creating global standards, Annex VI provides for the establishment of Emission Control Areas (ECAs). Stringent regulatory requirements exist in ECAs for the emission of SOx, particulate matter (PM) and NOx. Currently ECAs exist for the Baltic Sea (SOx), the North Sea (SOx) and North America (NOx and SOx). The area of the North American ECA extends up to 200 nautical miles from the coasts of the United States, Canada and the French territories and covers waters adjacent to the Pacific coast, the Atlantic/Gulf coast and the eight main Hawaiian Islands (EPA 2011).

4.3.1 Oxides of nitrogen emissions

Reductions in NOx emissions from marine diesel engines form part of Annex VI. The pollutant NOx is formed during the combustion process and NOx emissions from marine diesel engines contribute to ground level ozone, PM, eutrophication, acid deposition, nitrification and indirect effects that contribute to global warming. Regulation 13 of the 1997 Protocol and the NOx Technical Code apply to diesel engines of 130 kilowatt (kW) or more. According to Regulation 13, the NOx emissions of any diesel engine with a power output of more than 130 kW installed on a ship constructed on or after 1 January 2000 or one that underwent or undergoes a major conversion on or after that date shall not exceed certain set limits.[11] All ships of 400 gross tonnage (GT) or above, engaged in voyages to ports or offshore terminals under the jurisdiction of other Parties, shall be subject to surveys (initial, intermediate and periodical) and an International Air Pollution Prevention Certificate (IAPPC) shall be issued by the flag State Administration or a duly authorized body. Existing ships have to comply and thus be equipped with an IAPPC at the first scheduled dry-docking after entry into force of the amendments to Annex VI, but in no case later than three years after.

4.3.2 Oxides of sulphur content in fuel

Annex VI addresses SOx by setting progressively tighter limits for the sulphur content in fuels and increasingly stringent limits within the various ECAs. Thus, if a ship is within a SOx regulated ECA, it has to use fuel with a sulphur content not exceeding 1.5 per cent by mass, or an exhaust gas cleaning system or any other approved apparatus to limit its SOx emissions. The sulphur limit in SOx regulated ECAs must be 1.0 per cent from 1 July 2010 and must decrease to 0.1 per cent from 1 January 2015. As of 1 January 2012, the global sulphur cap is to be 3.5 per cent and is scheduled to decrease to 0.5 per cent from 1 January 2020. However, this decrease is subject to a feasibility review to be completed by the IMO no later than 2018, which will consider, among other matters, the availability of compliant fuel.

4.3.3 Ozone-depleting substances

Annex VI includes new MARPOL 73/78 regulations concerning the management of ODS including halons and chorofluorocarbons (CFCs). Annex VI also stipulates that new installations containing ODS are prohibited on all ships. The regulations additionally prohibit the deliberate release of other ODS such as freons and halons found in air conditioning, refrigeration and fire fighting systems.

4.3.4 Incineration

Annex VI prohibits the incineration on board ships of certain products such as contaminated packaging material and polychlorinated biphenyls (PCBs).

5. THE INTERNATIONAL REGULATION OF SHIP-SOURCED GREENHOUSE GAS EMISSIONS

The reduction of GHG emissions from ships has been the subject of much deliberation within the MEPC of the IMO, which is at the forefront of action to reduce the international shipping sector's GHG emissions. However, progress in the creation of legally binding rules to reduce the sector's GHG emissions has been difficult due in large part to differences in view within the Committee between its developed and developing State members. The major legal issues in dispute have been the role of developing States in the distribution of GHG reduction measures, and the compatibility of the reduction measures with other relevant rules of international law, particularly concerning the adoption of possible MBMs.

5.1 International Maritime Organization Studies and Policy for the Reduction of Greenhouse Gas Emissions

The first study by the IMO dealing with climate change and the reduction of the GHG emissions of the international shipping sector was commissioned as a result of Resolution 8 'CO2 Emissions from Ships', adopted at the Diplomatic Conference on Air Pollution held in September 1997 (1997 MARPOL 73/78 Conference). Resolution 8 requires the IMO to study ship-sourced GHG emissions as part of a global inventory of GHG emissions. More specifically, the Resolution invited the MEPC to consider feasible CO_2 reduction strategies for international shipping. The result of this was the *Study of Greenhouse Gas Emissions from Ships*, which utilized 1996 figures, and which was presented to the forty-fifth session of the MEPC in 2000 (MEPC 45).

A further Resolution, A.963(23), on 'IMO Policies and Practices related to the Reduction of Greenhouse Gas Emissions from Ships' was adopted by the twenty-third session of the IMO Assembly in December 2003. It recognized CO_2 as the main GHG emitted by ships and directed the MEPC to identify and develop technical and operational measures and MBMs to reduce GHG emissions from international shipping. It also requested the MEPC to develop a GHG emission index for ships and guidelines for the application of that index. In October 2006, the MEPC adopted a work plan for the development of technical and operational measures necessary for the reduction of the GHG emissions of international shipping. The first GHG study was updated by the 'Second IMO GHG Study 2009' (IMO 2009b). This second study addresses the principal GHGs (CO_2, CH_4, N_2O, HFCs, PFCs, SFs) and other relevant substances (NOx, NMVOC, CO, PM, SOx). The 2009 study shows that exhaust gas is the major source of GHG emissions from the international shipping sector. The objectives of this study were to assess the present and future GHG emissions from international shipping, the potential for reduction of these emissions through technology and policy and their impacts on climate change.

The current policy options for the reduction of ship-sourced GHG emissions that are being considered by the MEPC of the IMO comprise:

- mandatory limit on the Energy Efficiency Design Index (EEDI) for new ships
- mandatory or voluntary reporting of the EEDI for new ships
- mandatory or voluntary reporting of the Energy Efficiency Operational Indicator (EEOI)
- mandatory or voluntary use of a Ship Energy Efficiency Management Plan (SEEMP)
- mandatory limits on the EEOI value, combined with penalties for non-compliance
- a Maritime Emissions Trading Scheme (METS)
- an International Compensation Fund (ICF), to be financed by a levy on marine bunkers (IMO 2009b, p. 12; MEPC 2009).

5.2 Application of Greenhouse Gas Emissions Reduction Measures

All of the IMO's member States have publicly accepted the urgency of reducing GHG emissions in the international shipping sector. Many of the member States have also accepted the primacy of the IMO as the competent international organization within which to promulgate these technical rules, and Annex VI of MARPOL 73/78 as the appropriate regulatory instrument. However, beyond that fundamental consensus, attempts to make energy efficiency standards of design and operation mandatory have been resisted by a bloc of developing States within the IMO. Even more difficulty and controversy has surrounded the adoption of suitable MBMs.

In regulating the GHG emissions from international shipping, a rift emerged within the IMO between its developed and developing State members. At the fifty-eighth meeting of the MEPC in 2008 there was much debate concerning the application of measures, principles and policy. Developing States supported the application of the principle of CBDR as provided for by the UNFCCC, the Kyoto Protocol and the Copenhagen Accord, under which mandatory measures to reduce GHG emissions should be applicable exclusively to Annex I States. More specifically, these States opposed mandatory energy efficiency standards for ships based on CBDR and respective capabilities, and have argued that their economies would be disproportionately affected by any mandatory measures. They view the reduction of GHG emissions not solely as a climate change issue, but also as one of international development. Thus, the developing States have unanimously supported a position in which they should be allowed more time before making their GHG reduction measures mandatory.

Conversely, developed States have expressed the view that any IMO measures regarding safety of operations and protection of the marine environment from pollution, including the reduction of GHG emissions through international regulation, should operate irrespective of ship nationality. In other words, any IMO regulatory regime should be applicable to all ships engaged in international trade regardless of their flag, as the application of the principle of CBDR to international shipping does not operate effectively due to the complexity of the international shipping sector. In many instances, ships are registered in one State, while their beneficial owners are located in another State. Thus, international shipping does not suit inclusion within national emission targets. Further, the application of the principle of CBDR to developing States would cause significant 'carbon leakage'. In 2010, only about 35 per cent of the world merchant fleet was registered in Annex I States (ICS 2009). Developed States have maintained the position that a global regime allowing the international shipping sector to reduce its GHG emissions with significant environmental benefits requires the creation of a uniform reduction regime.

The United States presented a proposal to the sixtieth meeting of the MEPC in September 2010, requiring mandatory energy efficiency standards for ships and a system for trading in energy efficiency credits. However, these proposals could not be adopted because of disagreement by the bloc of developing States (MEPC 60/4/12). The sixty-first meeting in 2010 was the last opportunity for the IMO to demonstrate its ability to reduce the GHG emissions from international shipping before the Cancun meeting of the UNFCCC States Parties. In opening the sixty-first meeting, the IMO's outgoing Secretary-General, Efthimios Mitropoulos, emphasized that despite the Expert Group on MBMs not having reached conclusive recommendations, their work should be welcomed for its in-depth and prudent assessment of options. He also expressed the view that the IMO should continue to be the forum regulating the GHG emissions of international shipping and that Annex VI of MARPOL 73/78 was the appropriate legal instrument for implementing mandatory technical and operational measures to reduce GHG emissions, namely through the EEDI and the SEEMP (MEPC 2010). The Secretary-General further commented that the international shipping sector should not be considered an easy 'cash cow' for funding global reductions, and stressed that the sector's contribution to climate funding should be proportionate to its contribution to total global GHG emissions.

However, at this meeting, the divide between developed and developing State members increased, with the meeting failing to reach consensus on any of the major issues debated. Most of the delegates from developing States were of the view that mandatory measures should not be adopted for any one or more of the following reasons: firstly, the international shipping sector does not significantly impact on global warming; secondly, any IMO regulation places an unfair burden on developing States; thirdly, Annex VI of MARPOL 73/78 is not the appropriate regulatory instrument for reducing the GHG emissions from ships; and fourthly, because of a general support for the more extensive consideration of the position of developing States. The position of China in this regard is illustrative. While acknowledging the important contribution of the IMO to international shipping, China proposed a full application of the principle of CBDR as espoused by the UNFCCC and the Kyoto Protocol and thus lobbied the position that the IMO should strive for technical and operational solutions that fully comply with CBDR (MEPC, 61/24, Annex 3, p. 2).

More specifically, China objected to the inclusion of the EEDI, the SEEMP and the EEOI into Annex VI of MARPOL 73/78 for the following reasons: firstly, the technical measures are not yet adequately mature;

secondly, the measures do not embody the principle of CBDR; thirdly, CO_2 emissions and other GHGs are not air pollutants and therefore should not be included in the air pollution regulations contained in Annex VI of MARPOL 73/78; and fourthly, there are no clear provisions on the responsibilities of developed States in the provision of financial, technical and capacity building support to developing States (MEPC 61/24, Annex 3, p. 2). China instead made the following proposals: firstly, that the EEDI and the other technical and operational measures should remain voluntary in their application or that they should be made compulsory only for developed States; secondly, that if the EEDI and the SEEMP measures are to be made compulsory, they need to be created either within a new treaty or a new protocol to MARPOL 73/78 and contain clear provisions on the position of developing States; and thirdly, that the IMO should not hasten to adopt a METS, as ETSs are still being explored by the COP to the UNFCCC (MEPC 61/24, Annex 3, pp. 2–3). China's statements were echoed by India, and Brazil, Saudi Arabia, Qatar, Ecuador and Bolivia fully supported the views expressed by China. In an intervention by the UNFCCC Secretariat, it was stated that even though the IMO has contributed to the development of rules in the area: 'It is vital to synergize the capabilities and expertise of the IMO and the UNFCCC, and be cognizant of the progress under both processes, which are independent, Party driven, and distinct treaty regimes with their own Parties and sovereign decision-making bodies' (MEPC 61/24, Annex 6, p. 1).

The sixty-first meeting was also updated on negotiations at the UNFCCC for a global climate change agreement, which includes consideration of a global emissions cap on the maritime sector. There was a push from some States at the discussions of the UNFCCC for a levy on the international shipping sector to fund climate change adaptation and emissions reductions in developing States.[12]

Further talks on all three pillars of ship-sourced GHG reduction measures took place at the sixty-second meeting of the MEPC, which was held from 11 to 15 July 2011. A major breakthrough occurred at the meeting; namely, the adoption of mandatory technical and operational measures to reduce the GHG emissions of the international shipping sector. The MEPC also agreed to the Terms of Reference for an inter-sessional working group on energy efficiency measures for ships to take place during February/March 2012.

5.3 Technical and Operational Measures

The adoption of technical and operational measures combating the GHG emissions of the international shipping sector has been gathering momentum within the deliberations of the MEPC. Since the fifty-eighth meeting of the MEPC in 2008, there has been a steady development of the substantive technical and operational measures combating ship-sourced GHG emissions. At this meeting, agreement was reached to adapt the following terminology: 'Design CO_2 Index' became the 'EEDI', and 'Operational CO_2 Index' became the 'EEOI'. In 2009, at the fifty-ninth meeting, the Committee adopted the 'Second IMO GHG Study'. The momentum continued through to the sixty-first meeting in 2010, where within deliberations it was proposed that the Secretary-General circulate proposed amendments to Annex VI of MARPOL 73/78 requiring the EEDI and the SEEMP, both of which had previously been disseminated for voluntary use, to be made mandatory. Although mandatory measures were not adopted at the sixty-first meeting, considerable progress was made on all three fields of the Committee's GHG work. The meeting concluded that the Annex VI proposed amendments, to make the EEDI and the SEEMP mandatory, would be presented and discussed at the sixty-second meeting of the MEPC (from 11 to 15 July 2011) with a view to making them mandatory.

5.3.1 Technical measures: the Energy Efficiency Design Index

The EEDI is the most important technical measure for the reduction of the GHG emissions of international shipping. The adoption of the EEDI imposes new standards on both ship design and the lifting of fuel efficiency to reduce GHG emissions and the corresponding footprint of the international shipping sector. The EEDI will initially cover the largest emitters; namely, oil and gas tankers, bulk carriers, general cargo carriers and container ships. The EEDI of ships with lesser emissions will be addressed at an unspecified point in the future. Adopting the EEDI is critical, as it establishes the standards for ship design and building, as well as for emissions measurement and operating procedures. The purposes of the EEDI are: firstly, to require a minimum level of energy efficiency for new ships; secondly, to stimulate continued technical development of all components effecting the fuel efficiency of ships; thirdly, to separate the technical and design-based measures from operational- and commercial-based measures; and fourthly, to enable a comparison of the energy efficiency of individual ships to similar ships of the same size, which could have undertaken the same transport work and moved the same cargo (MEPC 59 Tech and OP Reduction Measures, para 2.2).

The EEDI is a non-prescriptive performance-based mechanism that identifies the required energy levels to be attained. It leaves the choice of technologies to be utilized in terms of specific ship design and specifications to industry. Ship designers and engineers are thus free to use the most cost-efficient solutions in ensuring a ship's compliance with the regulations. All that is prescribed is the required level of energy efficiency that needs to be attained. The reduction level set in the first phase will be 10 per cent. The EEDI will then be tightened incrementally every five years, driving the technical development of the fuel efficiency of ships. The IMO is anticipating reduction rates until the period 2025 to 2030, by which time a 30 per cent reduction rate will be mandated for most types of ships (MEPC 59 Tech and OP Reduction Measures, para 2.3).

The meetings of the MEPC have resulted in steady progress in the development of the EEDI for new ships, together with Interim Guidelines on the method of calculation of the EEDI for trial purposes, with a view to its further refinement and improvement.[13] The fifty-eighth meeting of the MEPC in 2007 invited delegations and observers to disseminate the Interim Guidelines to the broader maritime community to gauge their adequacy as a tool to improve energy efficiency for new ships. A mandatory limit on the EEDI for new ships appears to be a cost-effective solution and one that promotes improvements in the design efficiency of new ships. However, the EEDI only addresses the design of new ships, and this limitation is clearly a shortcoming that affects the usefulness of the EEDI. Further, it is important to note that the mandatory or voluntary reporting of the EEDI has no environmental effect *per se*, but rather environmental effectiveness depends on concurrently established incentive schemes (MEPC 2009). The sixtieth meeting of the MEPC in March 2010 demonstrated that the majority of State delegations supported Annex VI of MARPOL 73/78 as being the appropriate regulatory instrument within which to adopt mandatory technical and operational energy efficiency measures. Further, at this meeting, an in-principle agreement was reached that a Working Group would produce the more technical details of the EEDI, which were required before its implementation could occur.

The sixty-first meeting of the MEPC determined that work would continue on the Guidelines and on the methods of calculating the EEDI. Further, it was agreed that the Vice-Chair's Report on capacity building would be maintained. The implementation of the EEDI, a solution espoused by the United States, Japan, and others at the sixty-first meeting was widely criticized and thus no progress was made towards its adoption and inclusion in amendments to Annex VI of MARPOL 73/78. Even though the meeting did not secure the mandatory adoption of the EEDI, it accepted the need for an inter-sessional group to investigate further various issues surrounding the EEDI (MEPC 61/24, p. 27). In addition, a timeline of two years to finish determining EEDI baselines was agreed on by the MEPC.

At the sixty-second meeting of the MEPC in July 2011, the EEDI was finally resolved and agreement was reached that its use would be made mandatory. The EEDI regulations are contained in a new Chapter 4 of Annex VI of MARPOL 73/78; they will apply to all ships of over 400 GT and will commence operation on 1 January 2013 (IMO 2011b; Lloyd's 2010, p. 3). The new Chapter 4 regulations also include provisions on the promotion of technical cooperation, transfer of technology and exchange of information relating to improvement of the energy efficiency of ships. The regulations require State Administrations to cooperate with the IMO and other international organizations in promoting and providing support either directly or through the IMO to States requesting technical assistance, especially developing States in their implementation of measures to fulfil the requirements of their Chapter 4 obligations (IMO 2011b).

5.3.2 Operational measures: the Ship Energy Efficiency Management Plan and Energy Efficiency Operational Indicator

The complementary measure of the SEEMP, and how energy efficiency standards can be applied to existing ships, has also been under consideration by the MEPC. The SEEMP establishes a mechanism whereby shipping companies can improve the energy efficiency of their ship operations. The SEEMP will be ship-specific and should ideally be linked to the broader corporate energy management policy of the particular ship-owning corporation. A range of SEEMP models will ultimately be utilized, reflecting the wide variety of ship type, sizes and operating conditions. The SEEMP will provide an effective management tool that monitors a range of ship operations and facilitates the adoption of innovative technologies and practices, thereby optimizing the energy efficiency of ships. The SEEMP will operate through a continuous four-step cycle of ship energy-efficiency management, encompassing planning, implementation, monitoring and self-evaluation (MEPC.1/Circ. 683, 17 August 2009).[14] The use of the SEEMP is a feasible approach to increasing awareness of cost-effective measures to reduce GHG emissions among the IMO's member States. The cost effectiveness of the SEEMP will result in both less fuel consumption and a decrease in the environmental impacts of ship operations. The sixtieth meeting of the MEPC in 2010 further developed various fuel-efficiency management tools. For example, it finalized Guidance on the Best Practices for Fuel-efficient Operation of Ships, to be used in conjunction with the SEEMP under consideration.

The EEOI will be used in conjunction with the SEEMP and it will act as a monitoring tool and benchmark of the operations of different types of ship type and size. The mandatory limit of the EEOI is also a cost-effective solution that can provide a strong incentive to reduce GHG emissions. Progress has been made with the continued review of the EEOI and an inter-sessional correspondence group coordinated by Japan was established to advance the work further, to finalize it at the fifty-ninth meeting of the MEPC in July 2009. The MEPC now allows for voluntary use of the EEOI, to develop a consistent approach to the measurement of ships' energy efficiency, to assist in the evaluation of operational performance.

Despite the fifty-ninth and sixtieth meetings of the MEPC not being able to make the SEEMP mandatory under Annex VI of MARPOL 73/78, agreement was reached for a further review by the sixty-first meeting of the MEPC with a view to its ultimate adoption by the sixty-second meeting. It was also agreed that approval of the SEEMP by the flag administration (or its recognized organization) would not be required.

At the sixty-second meeting of the MEPC in July 2011, agreement was finally reached on the SEEMP. Provisions on the SEEMP will be included in the new Chapter 4 to Annex VI of MARPOL 73/78, requiring the mandatory adoption of the SEEMP for all ships. As with the EEDI, the provisions on the SEEMP will apply to all ships of over 400 GT and will commence operation on 1 January 2013 (IMO 2011b; Lloyd's 2010, p. 3). Further, the new Chapter 4 will contain the necessary definitions and requirements for survey and certification, including the prescribed format of the International Energy Efficiency Certificate (IMO 2011b).

5.3.3 Dealing with the legal uncertainty of the Energy Efficiency Design Index and the Ship Energy Efficiency Management Plan

For owners and brokers currently planning to sell and purchase new ships, it is recommended that the EEDI of the proposed new ship be included in any contract negotiations between the Parties. As for the SEEMP, it is recommended that such plans now be initiated to ensure that sufficient data is available for when the SEEMP becomes mandatory on 1 January 2013 (Lloyd's 2010, p. 3). Certain of the EEDI requirements that have not yet been finalized, such as safety consequences of speed reduction, ship size, ship types, target dates and reduction rates of ships with lesser air emissions, will continue to be worked on by the MEPC.

5.4 Market-based Measures

As global trade expands, significant growth is expected in international shipping, with a corresponding increase in the sector's GHG emissions. The MEPC has recognized at several of its meetings that technical and operational measures alone are not sufficient to reduce the quantity of GHG emissions of the sector satisfactorily. The adoption of MBMs would serve two main purposes: firstly, to allow an off-setting in other sectors experiencing a rise of ship emissions; and secondly, to provide an economic incentive for the shipping industry to invest in more fuel-efficient ships and technologies and to operate ships in a more energy efficient manner (MEPC 62, p. 4). The use of MBMs will thus assist the shipping sector in securing absolute reductions in its GHG emissions. Overall, economic or market-based mechanisms are favoured because of their place within the larger scheme of ETSs, such as those of the European Union and the UNFCCC/Kyoto Protocol. There is also a strong commitment within the IMO to establishing these market-based mechanisms. A METS and an ICF are cost-effective policy instruments with high environmental effectiveness, providing strong incentives for technological change, in terms of both operational technologies and ship design.

Within the range of MBMs currently being considered by the MEPC there are various versions of ETSs, efficiency trading systems and levies on bunker fuels. The ETSs under consideration will involve the emission of GHGs being subject to allowance quotas whereby rights to emit can be commercially traded through the ETSs. In this way, States and other entities that reduce their GHG emissions beyond their obligations can sell their excess emission capacity to those unable to meet their own targets. There are two main types of ETS that could be adopted: a 'cap and trade'[15] (Carbonpositive 2010) and a 'baseline and credit'[16] (Carbonpositive 2010). A METS would operate like other ETSs, such as those of the European Union and under the UNFCCC/Kyoto Protocol. An entity would be given an allowance, a limit up to which it can emit, and should it exceed its quota, it must purchase unused allowances to meet its prescribed target.

5.5 Progress Made on Market-based Measures

Considerable discussion has occurred within the MEPC on the adoption of appropriate MBMs. At the fifty-eighth meeting of the MEPC in 2008, there was consideration on an ETS, a global levy on fuel, and other hybrid market-based schemes for ships engaged in international trade. The majority of delegations opposed the development of any MBMs until all the issues surrounding the application of the principle CBDR were resolved in the UNFCCC context and in full recognition of Article 2.2 of the Kyoto Protocol. By contrast, other delegations expressed the view that as MBMs are a highly complex matter and still at a preliminary stage in their development, further information and studies were needed. The Committee decided to defer a focused debate on the matter to the fifty-ninth meeting of the MEPC. Thus, at the fifty-ninth meeting there was debate on offsets for the growing GHG emissions from ships, and incentives for industry investment in fuel and energy efficient shipping. Further, discussion took place on how the funds generated through MBMs should be utilized, such as for related purposes like mitigation and adaptation activities in developing States.

At the sixtieth meeting of the MEPC in 2010, agreement was reached on how to proceed in regards to MBMs, in which an MEPC Expert Group on Feasibility Study and Impact Assessment of Possible Market-Based Measures (Expert Group) would consider submitted documents and report to the sixty-first meeting of the MEPC. The Committee noted that it had identified several distinguishable MBM proposals and agreed that it needed to develop a methodology of feasibility studies and impact assessments to assess the proposed mechanisms. The Expert Group organized its work within a structure of four task groups: environment; shipping and maritime; administrative and legal; and trade, development and developing States. However, again, some developing States, in this instance Brazil, China, Cuba, India and Saudi Arabia, advocated for a postponement of any further work on MBMs until after the COP 16 of the UNFCCC.

The sixtieth meeting of the MEPC also agreed on a methodology, which is now contained in the Terms of Reference for the Expert Group. According to these Terms of Reference, the scope of the feasibility study and the impact assessments will review the practicability of implementing the various MBM options. Further, the study and assessment will also identify the reduction potential of GHG emissions and the impact on world trade, the shipping industry and the maritime sector, giving priority to the maritime sectors in developing States. Lastly, the study and assessment is to provide information on how the difference in socioeconomic capability between developing and developed States, and the special needs and circumstances of developing States, can be addressed (IMO 2010, Annex 2, Appendix 1, pp. 33–34).

The sixty-first meeting of the MEPC held extensive debate on how to progress on the development of suitable MBMs for international shipping. Substantial progress on such measures, however, was not expected at this meeting. The Committee instead discussed the way forward in the adoption of MBMs and the comprehensive Report of the Expert Group on feasibility and impact assessment of the several MBMs. In its Report, the Expert Group identified the proposals presented in Table 9.1.

The proposed MBMs considered by the Report utilized eight mechanisms for the reduction of GHG emissions, comprising both in-sector and out-of-sector mechanisms. However, the Expert Group found that the proposals lacked sufficient detail to varying degrees and that their evaluation was complicated by the different levels of maturity of the proposals. It concluded that it needed more information on all the possible measures that required further elaboration and development before being able to make any definitive policy assessment.

Table 9.1 Proposals identified by the MEPC Expert Group on Feasibility Study and Impact Assessment of Possible MBMs, as presented at MEPC 61

MEPC reference	Proposal made by	Proposal
MEPC 60/4/8	Cyprus, Denmark, the Marshall Islands, Nigeria and International Parcel Tankers Association (IPTA)	An International Fund for Greenhouse Gas Emissions from Ships (GHG Fund)
MEPC 60/4/37	Japan	Leveraged Incentive Scheme (LIS) to Improve the Energy Efficiency of Ships Based on the International GHG Fund
MEPC 60/4/40	Jamaica	Achieving Reduction in Greenhouse Gas Emissions from Ships through Port State Arrangements Utilizing the Ship Traffic, Energy and Environment Model, STEEM (PSL)
MEPC 60/4/12	United States	The United States Proposal to Reduce Greenhouse Gas Emissions from International Shipping, the Ship Efficiency and Credit Trading (SECT)
MEPC 60/4/39	World Shipping Council (WSC)	Vessel Efficiency System (VES)
MEPC 61/4/22	Norway	The Global ETS for International Shipping
MEPC 60/4/26	United Kingdom	Global ETS for International Shipping
MEPC 60/4/41	France	Further Elements for the Development of an ETS for International Shipping
MEPC 60/4/10	Bahamas	Market-Based Instruments: A Penalty on Trade and Development
MEPC 60/4/54	Germany	Impact Assessment of an ETS with a Particular View on Developing Countries
MEPC 60/4/55	International Union for Conservation of Nature (IUCN)	A Rebate Mechanism (RM) for a Market-Based Instrument for International Shipping

Source: IMO, Market-Based Measures Proposals Under Consideration within the Expert Group on Feasibility Study and Impact Assessment of Possible Market-Based Measures (MBM-EG) http://www.imo.org/OurWork/Environment/PollutionPrevention/AirPollution/Documents/Summary%20of%20MBM-EG%20proposals.pdf

At the sixty-first meeting, there were important interventions by India and China. The Indian delegation made the following statements:

- Firstly, apart from the UNFCCC issues, there are also questions of compatibility with the rules of the World Trade Organization (WTO). In particular, there is potential conflict with the non-discrimination principle and these issues have to be addressed before any conclusions can be drawn.
- Secondly, none of the MBMs are acceptable to India.
- Thirdly, the proposal to levy an additional charge on ships bunkers is not sound, as it amounts to the placing of a tax on international trade, a sector that is least responsible for global warming. Looking for a solution to global warming through international trade and shipping is therefore suspect.
- Fourthly, the effectiveness of an ETS in a global context is untried and unknown. Only the European Union has implemented one such system for its manufacturing sector. The experience is thus limited. It has been reported that in the European Union, ETS verified emissions decreased by around three per cent in 2008. Further, there are reports that suggest that the Union's ETS has generated multi-billion windfalls for some of Europe's biggest polluters. Given the limited experience, it is thus unclear whether this system would work in the international shipping sector. It has also been argued that the ETS would reward carbon efficiency improvements and the introduction of low-carbon technology, and that the ETS would provide financial incentive for the operation of ships to become more energy efficient. However, these benefits will apply only to those States that are in a position to invest in low-carbon ships.
- Fifthly, the Vessel Efficiency System (VES) focuses on improved fuel and carbon efficiency. Fuel charges would apply only if a ship fails to meet the applicable standard. This will fully eliminate the charge for ships meeting the efficiency standard and set a variable charge for ships failing to meet the standard. This proposal conflicts with international law, as it falls outside the scope of the UNFCCC and violates principle 11 of the Rio Declaration.
- Sixthly, the MBM proposals violate the principle of CBDR and respective capabilities based on equity. The application of these principles is now settled in international law. As most of the proposals negate these principles, they are thus not supportable.
- Seventhly, the proposals need to be consistent with the rules of the WTO and it is questionable whether they could easily comply with

the principle of non-discrimination as provided for by Article 1 of the General Agreement on Trade and Tariffs 1994 (GATT).

- Eighthly, climate change is a global problem and requires a global cooperative regulatory effort. There will be massive costs involved in stabilizing and reducing GHG emissions through the adoption of low-carbon technologies. These costs must be equitably shared among States on the UNFCCC principles of CBDR and respective capabilities. Any MBMs adopted must therefore be done so through the existing multilateral principles of climate change and be fair and equitable (MEPC 61/24, Annex 8, pp. 1–4).

At the sixty-second meeting of the MEPC in July 2011, the Committee considered the Report of the inter-sessional meeting of the Working Group that met in March 2011. The inter-sessional group held an extensive exchange of views on several issues relating to the desirability of MBMs, including such topics as the certainty of emission reductions or carbon price; revenues for mitigation, adaptation and capacity building activities in developing States; incentives for technical and operational improvements in international shipping; and off-setting opportunities.

5.6 The Way Forward on Market-based Measures

Important and complex questions remain unanswered on the various MBM proposals. These questions need to be resolved before there can be movement forward on any specific proposal. Many of these questions have already been identified by the Expert Group and have been the subject of discussion at the meetings of the MEPC. The questions can be grouped into three broad categories: choice of appropriate purpose and model; the relative position of shipping specific MBMs; and the compatibility of MBMs with other rules of international law.

Regarding choice of appropriate purpose and model, the questions are as follows: What is the specific need and purpose underpinning the adoption of MBMs? Should the particular MBMs under consideration reflect a partial reduction or a more absolute reduction of the GHG emissions of the international shipping sector? What is the model of MBM that should be adopted? What is more appropriate for the reduction of the GHG emissions of the international shipping sector? Is it an MBM that is subject to an explicit emission ceiling (a cap), or should it be a reduction target that comprises the entire world merchant fleet (a reduction target)? What should be the competent international organization for the reduction of the GHG emissions of the international shipping sector? Whether a cap or a reduction target is adopted, should the competent international organization within which these rules are created be the IMO or the UNFCCC? What are the criteria on which the cap or target would be established? What is the appropriate economic methodology with which the cap or target should be set and maintained? What kind of reduction target should be adopted? There are variations on the mode of expression of the target that could be adopted. The first option is a target for the entire world fleet that is expressed as a maximum amount of CO_2 emissions per year by an established target year, while a second option would be an annual percentage reduction from a fixed year onwards. Both methods require an agreed base year from which calculations are made. Further, should efficiency improvement targets be set for the entire fleet or should separate targets for each part of the fleet be used? For example, should the same efficiency improvement targets be set for bulk carriers and very large commercial transporters (VLCTs), for which energy efficiency is expressed as per capacity mile? Finally, how much further evaluation of MBMs is needed? The COP to the UNFCCC is still exploring various ETS models. The IMO is also further evaluating proposed MBMs that are currently being considered by the Expert Group, including regarding their impact on international trade, the maritime sector of developing States, least-developed countries (LDCs) and small-island developing states (SIDS), and the corresponding environmental benefits.

On the topic of the relative position of shipping specific MBMs, the questions are as follows. Will other modes of international transport be the subject of regulation similar to that of the international shipping sector? What will the two other transport modes – namely, international civil aviation and road transport – do? How will their GHG emissions be regulated internationally and will MBMs be adopted in respect of their GHG emissions? The inclusion of both the international civil aviation and shipping sectors within the European Union's ETS for GHG emissions has been proposed by a number of European Union member States. Examination is needed on how emissions trading can be adapted for use by the diverse modes of global transport. What quantity of GHG emissions should be allocated to the international shipping sector? How much possible future 'carbon space' should be allowed for the international shipping sector, a cost-effective and energy-efficient industry that transports 90 per cent of world trade and the sustainably of which underpins the global economy?

Finally, the following questions remain to be answered regarding the compatibility of MBMs with other rules of international law. How can the interests of the two distinct multilateral regimes of the IMO and the UNFCCC be reconciled? There are significant compatibility issues between the work of the IMO in creating MBMs and the work of the

UNFCCC. On the one hand, the UNFCCC, the Kyoto Protocol and the Copenhagen Accord advocate the interests of developing States and the rationalization of the principle of CBDR; with relative capabilities thereby placing a greater share of GHG reduction responsibility on developed States for current and projected climate effects. On the other hand, there is no precedent in any of the IMO's 52 multilateral instruments of measures being applied selectively to certain ships. Quite to the contrary, the IMO's regulatory measures apply to all ships regardless of their place of registration, or their flag and nationality. As was noted by the UNFCCC Secretariat, even though the IMO has contributed to the development of rules in this area, it must synthesize its capabilities and expertise along with the UNFCCC, and be cognizant of the progress under both processes. What are the compatibility issues of possible MBMs with relevant rules of the WTO? This is important because these compatibility issues are common, if experience of conflict between the principle of non-discrimination and other rules of the WTO with multilateral environmental agreements are any indication. Are there compatibility issues between possible MBMs and the customary international law of the sea as reflected in the LOSC? What will be the impact of possible MBMs on international trade and sustainable development, including their impact on the maritime sector of developing States, LDCs and SIDS? What are the corresponding environmental benefits?

6. CONCLUSION

International shipping underpins global commerce and the global economy, while at the same time, sea transport is predicted to continue rising consistent with world trade. It is thus imperative that the international shipping sector reduces its GHG emissions. To this end, the IMO has been actively developing rules for the reduction of the GHG emissions of international shipping. In 1997, the Organization adopted Resolution 8 of the 1997 Conference of Parties to MARPOL 73/78 to examine the CO_2 emissions from international shipping. Exhaust gas in the form of CO_2 was identified as the main GHG emitted by ships and the MEPC was directed to identify and develop technical and operational measures, as well as MBMs, to reduce the GHG emissions of the sector.

In promulgating rules concerning technical, operational and market measures, a rift emerged within the Committee between its developed and developing member States. The role of developing States in terms of their responsibility to reduce their GHG emissions has been the primary issue in dispute. They have strongly relied on the application of the principle of

CBDR to the reduction of the GHG emissions and have questioned the appropriateness of applying uniform international standards, which they view as economically disadvantageous. The IMO ought to be viewed as the competent international organization within which to develop rules for international shipping to reduce GHG emissions. The appropriateness of the IMO to promulgate these rules is supported by the LOSC, the Kyoto Protocol, the Copenhagen Accord and Agenda 21. It follows that Annex VI of MARPOL 73/78 is the correct regulatory instrument for the reduction of ship-source GHG emissions.

The adoption and implementation of energy efficiency standards through the EEDI and the SEEMP as well as through associated programmes such as the EEOI is critical, as it establishes the standards for ship design, emissions measurement, operating procedures, and monitoring and evaluating mechanisms. Major progress has been made by the MEPC in creating rules concerning both technical and operational measures. Until the sixty-second meeting of the MEPC, the use of both the EEDI and the SEEMP, which had been disseminated through MEPC Circulars, was voluntary only. However, at the sixty-second meeting, the necessary consensus was reached to make the use of the EEDI and the SEEMP mandatory through the adoption of measures contained in Chapter 4 to Annex VI of MARPOL 73/78. The international shipping sector is now the first international industry sector to adopt a mandatory GHG reduction scheme. If the EEDI is put in place by 1 January 2013 as expected, it is anticipated that by 2020 some 45 to 50 million tonnes of CO_2 will have been removed from the atmosphere. As for the SEEMP, the operational efficiencies that will be gained will make a valuable contribution to reducing global CO_2 emissions.

Less progress has been made with respect to MBMs, which will require further work before an appropriate model can be adopted. There are numerous MBM proposals currently being considered by the Expert Group of the MEPC. The comprehensive Report of the Expert Group provided an opinion on the compelling need and purpose of MBMs and studied the feasibility and impact assessment of the various MBM or carbon price measures. The scope of the work of the Expert Group was to evaluate current proposals with the aim of assessing the extent to which they could reduce the GHG emissions from international shipping while giving priority to the maritime sectors of developing States, LDCs and SIDS. Further, there are complex legal issues involved concerning the compatibility of any proposed MBMs with other rules of international law, particularly the rules of the UNFCCC, Kyoto Protocol and Rio Declaration, and the rules of the WTO and customary international law as reflected in the LOSC. The member States of the IMO need to systematically resolve these issues, including the compatibility issues, and incorporate all relevant rules of international law into any MBM proposals that are under consideration before any real progress can be made. These issues will not be easy to resolve and an MBM should not be adopted hastily.

NOTES

1. The impact of increased greenhouse gas (GHG) concentrations in the atmosphere results in rising global temperatures and climatic change. The effect of this includes rises in sea levels caused by the melting of polar ice caps and glaciers, as well as more frequent and intense weather phenomena. Climate change has direct effects on the marine environment, including damage to coral reefs; acidification of the sea; increased storm activity; adverse impacts on fisheries; melting of sea ice; and changes to wave conditions.
2. Through Annex VI of the International Convention for the Prevention of Pollution from Ships (MARPOL 73/78), the International Maritime Organization (IMO) has adopted laws regulating ships' emissions of Oxides of Nitrogen (NOx), the levels of Oxides of Sulphur content in fuel (SOx), the deliberate emissions of ozone depleting substances, and incineration.
3. The views of the IMO's member States vary on the adoption of mandatory measures. Developing States generally oppose mandatory energy efficiency standards for ships based on the principle of common but differentiated responsibility (CBDR) as pronounced in the United Nations Framework Convention on Climate Change (UNFCCC) and the Kyoto Protocol to the UNFCCC (Kyoto Protocol). In particular, developing States argue that if differentiating and equitable principles are not applied, their economies will be disproportionately affected by the adoption of mandatory measures. The application of the principle of CBDR to international shipping is a difficult issue. On the one hand, the UNFCCC and the other climate change instruments support the application of CBDR, while on the other, because of the necessity of applying uniform international shipping standards as well as the nuances of ship nationality, the application of CBDR does not operate smoothly.
4. There are varying views on the meaning and content of Sustainable Development. Sustainable Development can be seen as having an economic, social and environmental dimension. The present discussion centres on the concept's environmental dimension and the constituent environmental principles of Sustainable Development.
5. Annex I States are all States members of the Organization for Economic Cooperation and Development (OECD) and economies in transition (EIT). Annex II States are OECD member States, but are not included in the EIT.
6. The United States has signed but not ratified the Kyoto Protocol. Australia ratified the Kyoto Protocol on 3 December 2007.
7. Assigned amount is the percentage in Annex B x aggregate emissions in 1990 x 5.
8. The IMO became active when its constituent treaty entered into force in 1958. Its original name was the Inter-Governmental Consultative Organization (IMCO) and its name was changed to the IMO in 1982. It currently has 167 member States and three associate members.
9. Chapter 17 provides an action programme for the achievement of: integrated management and sustainable development of coastal areas, including exclusive economic zones; marine environmental protection; sustainable use and conservation of marine living resources of the high seas; sustainable use and conservation of marine living resources under national jurisdiction; addressing critical uncertainties for the management of

the marine environment and climate change; strengthening international, including regional, cooperation and coordination; and sustainable development of small islands.

10. The shipping sector has been seeking alternative methods for supplying energy to ships while in ports. When in port, ships need significant amounts of energy to power lights, heating, hot water, air conditioning, and refrigerated cargo, while for cruise liners there are increased demands on energy. The energy for these activities has traditionally been supplied internally; that is, through the ships' diesel engines. The shipping sector has for some time now been looking at alternative energy sources that are more environmentally friendly. One proposal that is receiving a lot of attention from the shipping sector and ports around the world is 'alternative maritime power' (AMP), also referred to as 'cold ironing' or 'shore power'. The use of AMP requires ships to turn their diesel engines off when docked in port and to 'plug-in' to the local energy grid, rather than using their diesel auxiliary engines. This replaces 'dirty' shipboard energy sources with more environmentally friendly ones. When a ship's diesel engines are not operating, its emissions are greatly reduced. With the ever-increasing attention on pollution reduction, there are several factors in support of the use of AMP. The benefits of AMP are: firstly, a significant reduction of ships' emissions while in ports, in particular of NOx, SOx and PM; secondly, a reduction of emissions of the GHG NOx; thirdly, the minimization of health risks and associated adverse impacts on port workers, on-board personnel and the inhabitants of ports communities; and fourthly, a reduction in the cost of energy, as the cost of producing electricity on ships when they are in port is much higher than the cost of purchasing energy produced on shore. Despite the obvious benefits in the use of AMP, there are some reservations about its widespread use. Firstly, AMP is feasible only if the shore-based power production facility uses environmentally friendly renewable resources; and secondly, feasibility is reliant on whether the shore-based facility has sufficient grid power to accommodate the widespread use of AMP across major ports.
11. 17 g/kW-hr when *n* is less than 130 rpm 45.0**n*(-0.2)g/kW-hr when *n* is 130 or more, but less than 2000 rpm 9.8 g/kW-hr when *n* is 2000 rpm or more.
12. The United States-Caribbean Emissions Control Area (ECA) that would have affected the United States Virgin Islands and Puerto Rico economies was also rejected.
13. The purpose of the Interim Guidelines on voluntary verification of the Energy Efficiency Design Index (EEDI) was to assist verifiers of the EEDI in carrying out verification in a consistent manner that would also assist the MEPC with the ultimate adoption of mandatory measures. The Interim Guidelines were also intended to assist shipowners, shipbuilders and engine and equipment manufacturers with their understanding of the EEDI verification procedures (MEPC 62, Interim Guidelines for Voluntary Verification of the Energy Efficiency Design Index http://www.mardep.gov.hk/en/msnote/pdf/msin0923anx2.pdf).
14. A Ship Energy Efficiency Management Plan (SEEMP) should include best practice on the following maritime components: voyage performance; optimized ship handling; hull and propulsion system maintenance; use of a waste heat recovery systems; improved fleet management; improved cargo handling; improved energy management; fuel types; compatibility of measures; age; operational service life; and trade and sailing area (MEPC 61, Guidance for the Development of a Ship Energy Efficiency Management Plan (SEEMP) http://www.imo.org/blast/blastDataHelper.asp?data_id=26530&filename=683.pdf pages 5-10).
15. Cap and trade is the most popular emission-trading scheme (ETS), whereby emissions are subject to a cap, permits are issued up to that cap, and the market allows those emitting less than their quotas to sell their excess permits to emitters needing to purchase extra permits to meet their own reduction quotas.
16. Baseline and credit is a type of ETS in which entities are required to reduce their GHG emissions below a projected 'business as usual' path of increasing emissions. Any reduction below the future path earns credits for the difference that can be sold to other entities unable to contain increases to baseline levels.

REFERENCES

Agenda 21, accessed May 2011 at: http://www.un.org/esa/sustdev/documents/agenda21/english/agenda21toc.htm.

Bright Hub (2011), accessed May 2011 at: http://www.brighthub.com/engineering/marine/articles/24522.aspx.

Cameron J. and J. Abouchar (1991), 'The Precautionary Principle: A Fundamental Principle of Law and Policy for the Protection of the Global Environment', 14 *Boston College International & Comparative Law Review* 1.

Carbonpositive (2010), *Glossary*, accessed May 2011 at: http://www.carbonpositive.net/viewarticle.aspx?articleID=44.

EPA (United States Environmental Protection Agency) (2011), *Designation of North American Emission Control Area to Reduce Emissions from Ships: Regulatory Announcement*, accessed May 2011 at: http://www.epa.gov/nonroad/marine/ci/420f10015.htm.

Franckx, E. (1997), 'The ILA Survey on Coastal State Jurisdiction with Special Reference to Regional Rules', in H. Ringbom (ed.) *Competing Norms in the Law of Marine Environmental Protection* (Kluwer Law International).

ICS (International Chamber of Shipping), *IMO's Track Record on Environmental Regulation*, accessed May 2011 at: http://www.shippingandco2.org/imotrackrecord.htm.

ICS (2009), *Shipping, World Trade and the Reduction of CO2 Emissions*, accessed May 2011 at: http://www.shippingandco2.org/CO2%20Flyer.pdf, at 3–4.

IMO (International Maritime Organization) (2005), *International Shipping: Carrier of World Trade*, accessed May 2011 at: http://www.imo.org/KnowledgeCentre/ShipsAndShippingFactsAndFigures/TheRoleandImportanceofInternationalShipping/IMO_Brochures/Documents/Internatioinal%20Shipping%20-%20Carrier%20of%20world%20trade.pdf, at 1.

IMO (2009a), *International Shipping and World Trade Facts and Figures*, accessed May 2011 at: http://www.imo.org/blast/blastDataHelper.asp?data_id=13865&filename=InternationalShippingandWorldTrade-factsandfigures.pdf, at 32.

IMO (2009b), The *Second IMO GHG Study 2009*, document, MEPC/59/INF.10.doc, 12, accessed May 2011 at: http://www.imo.org/OurWork/Environment/PollutionPrevention/AirPollution/Documents/INF-2.pdf.

IMO (2010a), *Cancún Climate Change Conference – IMO briefing*, accessed May 2011 at: http://www.imo.org/mediacentre/pressbriefings/pages/climate-change-conference.aspx.

IMO (2010b), *Marine Environment Protection Committee (MEPC) 61st Session: 27 September to 1 October 2010*, accessed May 2011 at: http://www.imo.org/MediaCentre/MeetingSummaries/MEPC/Pages/MEPC-61st-Session.aspx.

IMO (2011a), *Air Pollution and Greenhouse Gas Emissions*, accessed May 2011 at: http://www.imo.org/OurWork/Environment/PollutionPrevention/AirPollution/Pages/Default.aspx.

IMO (2011b), *Press Briefing*, accessed May 2011: http://www.imo.org/MediaCentre/PressBriefings/Pages/42-mepc-ghg.aspx.

LOSC, United Nations Convention on the Law of the Sea, Dec 10, 1982 UN Doc A/Conf 62/122 (1982), reprinted at 211 *ILM* 1261 (1982) (LOSC), entered into force on 14 November 1994.

Lloyd's (Lloyd's Register Briefing) (2010), *Marine Environmental Protection*

Committee IMO MEPC – March 2010: Outcome and Opportunities, accessed May 2011: http://www.lr.org/Images/IMO%20Marine%20Environmental%20Protection%20Committee%20-%20March%202010_tcm155-193338.pdf.

MARPOL 73/78, International Convention for the Prevention of Pollution by Oil from Ships (MARPOL 73) (London), 12 *ILM* (1973), 1319 (Amended by Protocol of 1978); Protocol Relating to the Convention for the Prevention of Pollution by Oil from Ships (MARPOL 78) (London), 17 *ILM* 546 (1978).

MEPC 45, Document MEPC 45/8.

MEPC/Circ.471, 2005.

MEPC 58, MEPC 58/INF.21, 1 August 2008.

MEPC 59, MEPC 59 1/Circ 684, 17 August 2009.

MEPC 59, Description of the Package of Technical and Operational Reduction Measures for Ships Agreed by MEPC 59, accessed August 2012 at: http://www.imo.org/OurWork/Environment/PollutionPrevention/AirPollution/Documents/Technical%20and%20operational%20reduction%20measures.pdf.

MEPC 60, MEPC 60/4/12, 14 January 2010.

MEPC 60, 60/22, 12 April 2010.

MEPC 60, 60/22/Add.1, 17 June 2010.

MEPC 61, 27 September to 1 October 2010 .

Address of the Secretary-General at the Opening of the Sixty-first Session of the Marine Environment Protection Committee (27 September to 1 October 2010), accessed August 2012 at: http://www.imo.org/mediacentre/secretarygeneral/secretary-generalsspeechestomeetings/pages/marine-environment-protection-committee-(mepc)-61st-session-27-september-to-1-october-2010--.aspx.

MEPC 61, Report of the Marine Environment Protection Committee on its Sixty-first Session, accessed August 2012 at: http://www.uscg.mil/imo/mepc/docs/mepc61-report.pdf.

MEPC 61, MEPC 61/5/3, 7 July 2010.

MEPC 61, MEPC 61/INF.12, 15 July 2010.

MEPC 61, MEPC 61/INF.2, 13 August 2010.

Palassis S. (2011), 'Beyond the Global Summits: Reflecting on the Environmental Principles of Sustainable Development', 22 *Colorado Journal of International Environmental Law and Policy* 41.

MEPC 61, MEPC 61/24, Report of the Marine Environment Protection Committee on its Sixty-first Session, accessed August 2012 at: http://www.uscg.mil/imo/mepc/docs/mepc61-report.pdf.

MEPC 61, MEPC.1/Circ.683, 17 August 2009, Guidance for the Development of a Ship Energy Efficiency Management Plan (SEEMP), accessed August 2012 at: http://www.imo.org/blast/blastDataHelper.asp?data_id=26530&filename=683.pdf page 2.

MEPC 62, Market-Based Measures, accessed August 2012 at: http://www.imo.org/ourwork/environment/pollutionprevention/airpollution/pages/market-based-measures.aspx.

Rio Declaration on Environment and Development UN Doc. A/CONF.151/26/Rev.1, *Report of the UNCED*, Vol. 1(New York); reprinted in 31 ILM 874 (1992).

UNFCCC (2006), Doc. Decision 10/CMP 2, FCCC/KP/CMP/2006/10/Add.1, accessed May 2011 at: http://unfccc.int/resource/docs/2006/cmp2/eng/10a01.pdf#page=36.

UNFCCC (2011a), United Nations Framework Convention on Climate Change, May 9, 1992, 31 ILM 849 (1992), entered into force on 21 March 1994.

UNFCCC (2011b), Kyoto Protocol to the UNFCCC (Kyoto Protocol), FCCC/CP/1997/C.7/Add1, 37 ILM 22, accessed May 2011 at: http://unfccc.int/resource/docs/convkp/kpeng.pdf, entered into force on 16 February 2005.
UNFCCC (2011c), *Status of Ratification of the Kyoto Protocol*, accessed May 2011 at: http://unfccc.int/kyoto_protocol/status_of_ratification/items/2613.php.
UNFCCC (2011d), *Kyoto Protocol*, accessed May 2011 at: http://unfccc.int/kyoto_protocol/items/2830.php.

10. Climate change mitigation activities in the ocean: turning up the regulatory heat

Rosemary Rayfuse and Robin Warner

1. INTRODUCTION

The adverse impacts of anthropogenically induced climate change on the terrestrial and marine environments have been acknowledged by a succession of expert reports commissioned by global and national bodies (IPCC 2007; Preston and Jones 2006; Stern et al. 2006). The threats posed by climate change to the global environment have fostered heightened scientific and commercial interest in a range of CO_2 sequestration methods that either involve the ocean or affect the marine environment. The most developed proposals to date relate to offshore carbon capture and storage (OCCS), which seeks to capture carbon dioxide from point sources of emissions and sequester it in sub-seabed geological formations. Considerable financial and technological investment has already been made in this approach, and a regulatory framework has been developed for its implementation both at the global and at the national level in Australia. Other methods seek to boost the capacity of the oceans to capture and absorb atmospheric carbon dioxide either through the deposit of substances or wastes into the ocean, or through the deposit of structures or devices into the ocean, to increase the production of organic material in the surface ocean and thereby promote increased draw down of photosynthesized carbon to the deep ocean. Proposals for these geo-engineering schemes include seabed deposition of biochar, increasing ocean alkalinity, increasing carbon concentrations in down-welling water, and using wave activated pumps to alter water circulation (Scientific Group of the London Protocol 2010). Most advanced are proposals for ocean fertilization, involving the deposition of iron, nitrogen or phosphorous compounds into the water column to stimulate primary productivity and increase carbon-flux to the deep ocean.

Both OCCS and marine geo-engineering schemes have the potential for as yet unknown and possibly adverse effects on the marine environment. Increasingly commentators have called for an assessment of these schemes from practical, political, social, ethical and legal perspectives (Lin 2009; Schneider 2008; Verlaan 2009). This chapter focuses on the legal challenges posed by OCCS and by marine geo-engineering methods, in particular ocean fertilization, and the progress that has been made in regulating these activities at the global and national levels. Section 2 describes these processes, their objectives and their potential impacts on the marine environment. Section 3 analyses the applicability of existing international law principles to these processes and the regulatory gaps and ambiguities in the existing international law framework for such activities. Section 4 examines the emerging policy and regulatory framework for these processes at the global and national level in Australia. Finally, in Section 5, it will be concluded that a significant contrast can be seen between the regulatory framework applied to OCCS and that applied, thus far, to ocean fertilization.

2. SEQUESTERING CARBON DIOXIDE IN THE OCEANS

2.1 Sequestering Carbon Dioxide in the Seabed: Offshore Carbon Capture and Storage

Carbon dioxide (CO_2) capture and storage (CCS) involves the separation of CO_2 from industrial- and energy-related sources and its transportation to and permanent storage in geological formations such as exhausted oil and gas fields, un-mineable coal beds and deep saline formations (IPCC 2007). Application of CCS technologies can thus allow continued use of fossil fuels and other CO_2 producing processes, while simultaneously lowering CO_2 emissions. In 2005, the Intergovernmental Panel on Climate Change (IPCC) estimated that CCS could potentially contribute between 15–55 per cent of cumulative emissions reductions worldwide to 2100 (GCCSI 2009). More recently, the International Energy Agency (IEA) estimated that it could potentially contribute 19 per cent of emissions reductions by 2050 (IEA 2008).

The concept underlying CCS is not new. Direct injection of CO_2 into subsurface geological formations has been used as part of enhanced oil recovery (EOR) projects since the early 1970s (IPCC 2007, p. 199). Since the 1990s, the technology has been used to capture and store CO_2 from natural gas processing operations. In 1996, following the introduction by the Norwegian Government of a tax on CO_2 emissions, the Norwegian state oil company, Statoil, opened the world's first offshore commercial CO_2 storage site at its Sleipner oil field. In recent years, consideration of the application of CCS technologies to pre- and post-combustion processes in power generation and other industrial processes has gained considerable momentum. According to the Global CCS Institute, there are currently 213 active or planned CCS projects globally. Of these, 101 projects are considered to be of commercial scale, proposing to sequester at least 1 Mtpa of CO_2. Of these commercial-scale projects, 62 are defined as 'integrated' projects, meaning that the capture, transport and storage is all undertaken by a single project owner or operator, thereby providing a 'full source to sink CCS solution' (GCCSI 2009, p. 9). To date, however, only seven of these projects are actually in operation (GCCSI 2009, p. 22).

The attractiveness of CCS is in part due to the potential for isolation of storage sites from population centres – particularly in offshore areas – and the perceived lack of potential for interaction with and damage to the surrounding environment. The potentially large number of storage sites, as well as the potential for adaptation of pre-existing technology and infrastructure within the oil and gas sector and other industries, has also contributed to its attractiveness as a key mitigation technology. Storage sites envisaged and already being used for CCS include both onshore and offshore depleted oil and gas fields and deep subterranean and sub-sea saline aquifers (Brewster Weeks 2007, p. 252; Scott 2005, p. 60). While, to date, offshore CCS, or OCCS, projects have taken place in waters within national jurisdiction, the potential also exists for future storage in seabed areas beyond national jurisdiction when storage sites within national jurisdiction are exhausted (International Maritime Organization – IMO 2007a).

Opponents of CCS point to a range of technological, financial, safety and regulatory issues that need to be resolved before CCS should be pursued. In the context of OCCS, the principal risk is the potential for leakage of carbon dioxide and any other substances in the carbon dioxide stream, such as heavy metals, into the marine environment, either during transport to a storage site or after storage (Abanedes et al. 2005, p. 18). Leakage of these substances into the marine environment can alter the marine chemistry of the water column and lead to adverse effects on the interconnected web of marine species, habitats and ecosystems (Abanedes et al. 2005, p. 18; Koslow 2007, p. 160). Considerable research has been undertaken by States in conjunction with corporations to assess and minimize the environmental risks associated with sub-seabed sequestration of carbon dioxide. Nevertheless, information about its long-term effects on the marine environment is lacking (Brewster Weeks 2007, pp. 252–253; GCCSI 2009, p. 9; Scott 2005, p. 60).

In 2005, the IPCC concluded that technological, cost and regulatory issues relating to regulation of site selection, capture, transport, oversight, monitoring, remediation and liability remained valid. To meet these concerns, in 2008 the G8 committed to supporting the launch of 20 large-scale CCS demonstration projects globally by 2010, to support technology development and cost reduction, with a view to enabling the broad deployment of CCS by 2020. While the United States, Europe, Australia and Canada all have programmes in place to support project development, this goal has not yet been met. Nevertheless, Australia has made the development of CCS a strategic priority (CGA) and has committed $100 million per annum to fund the Global CCS Institute, the central objective of which is to accelerate the worldwide development and implementation of commercial-scale CCS projects (GCCSI 2010).

2.2 Increasing Ocean Absorption of CO2: Ocean Fertilization

Augmenting the rate at which the oceans absorb carbon dioxide is the fundamental objective of ocean fertilization as a climate mitigation activity. Fertilization is designed to increase phytoplankton primary productivity in iron and other nutrient deficient areas of the ocean, thereby increasing the amount of 'marine snow' or organic detritus falling from the upper layers of the water column to the deep ocean (Koslow 2007, pp. 157–158; Rayfuse et al. 2008, pp. 302–303). In theory, carbon transported as marine snow into the deep ocean and finally decomposed to inorganic nutrients and dissolved carbon dioxide will remain out of contact with the surface ocean and atmosphere for the relatively long time scales associated with ocean currents and circulation (Cullen and Boyd 2008, p. 296).

At least four distinct fertilization methods have been proposed to date. The first, and most studied, involves fertilization by the deposition of large quantities of iron directly into the water column to enhance macronutrient uptake and local productivity. This technique has been proposed for use in the approximately 25 per cent of the world's ocean surface predominantly located in the Southern Ocean, where high concentrations of macronutrients exist, but where chlorophyll (plant biomass) concentrations are low. Production in these high-nutrient low-chlorophyll (HNLC) waters is primarily limited by micronutrient supply; in particular, the supply of iron (Lampitt et al. 2008). Thirteen artificial iron fertilization experiments have been conducted since 1993; the majority of them being conducted in the Southern Ocean, with limited and inconclusive results (Aumont and Bopp 2006; de Baar et al. 2005). Early experimental findings suggested that natural iron fertilization in HNLC regions promoted carbon export and sequestration by potentially measurable amounts. However, while the first 12 experiments clearly evidenced enhanced phytoplankton production (Boyd et al. 2007, p. 612; Buesseler and Boyd 2003, pp. 67–68; Koslow 2007, p. 159; Lampitt et al. 2008; Rayfuse et al. 2008, p. 305), they were not all designed to measure carbon export from the upper ocean and none of them was designed to measure sequestration (Lampitt et al. 2008, p. 3929). Preliminary conclusions drawn from the results of the most recent LOHAFEX ocean iron fertilization experiment conducted in 2009 indicated that the CO_2 drawdown effect of ocean iron fertilization was low (IMO 2010b, para. 2.13; Alfred Wegener Institute 2009). The efficacy of this technique in sequestering anthropogenic carbon has therefore been questioned (Lampitt et al. 2008, p. 3930).

A second method, similar to the first, involves the introduction of micro- or macronutrients such as iron, nitrogen and phosphorous into the water column in low-nutrient low-chlorophyll (LNLC) areas to enhance nitrogen fixation, thereby increasing primary production. Approximately 40 per cent of the ocean surface, located mostly in the subtropical gyre systems, is considered to be LNLC. However, little is known about the relationship between iron supply and limiting factors such as phosphorous supply, or about the mechanisms of nutrient supply in these areas. The efficacy of this method in enhancing sequestration similarly remains unstudied (Lampitt et al. 2008, p. 3930).

These first two methods involve the supply of nutrients from ocean-based sources; in other words, from a ship. A third method involves the supply of nutrients from land-based sources. In this scenario, the fertilizing nutrients are manufactured on land and transported by submarine pipe to the deposition site. Based on comparisons with leakage of agricultural fertilizer to coastal areas, it has been hypothesized that choice of both fertilizing agent and location of injection can be used to control for local conditions, thereby maximizing sequestration potential. However, the costs of producing the fertilizing agent and piping it from land are likely to be large, with a carbon footprint that may be greater than the carbon sequestered. In addition, the sequestration potential will be limited by the topography of the sea shelf, ocean circulation, local physics of the water column and general ecosystem dynamics and the carbon cycle, factors about which little is currently known. While initial studies costing the injection of urea, extracted from atmospheric nitrogen, suggested this was a viable proposition, the assumptions relating to nutrient supply at the injection site on which these calculations were based have been found to be incorrect, and any sequestration is likely to be only short-term and localized (Lampitt et al. 2008, pp. 3926–3927).

A fourth method involves the use of local wave power to pump deep nutrient rich water from depths of several hundred metres to the surface to enhance primary production and CO_2 sequestration. Studies have disputed the efficacy of this proposed method, noting that deep ocean waters contain elevated concentrations of dissolved CO_2, which may simply be released into the atmosphere when these deep waters reach the surface. Experimental results have been disappointing (Lampitt et al. 2008, pp. 3927–3928).

As already noted, a variety of risks and uncertainties associated with ocean fertilization have been identified by scientists and environmentalists. The effects of stimulating phytoplankton productivity on other marine organisms and marine ecosystems generally, is poorly understood (Koslow 2007, p. 159; Rayfuse et al. 2008, pp. 305–306; Scott 2005, pp. 87–88). In addition, the sinking of phytoplankton blooms into the deep ocean may reduce oxygen levels at these depths, leading to eutrophication and anoxia, with adverse consequences for fisheries and other marine organisms (Johnston et al. 1999, pp. 24–25; Lampitt et al. 2008, pp. 3926–3927; Rayfuse et al. 2008, p. 307). Increased productivity of phytoplankton may also boost the production of other greenhouse gases such as nitrous oxide, neutralizing the positive effects of enhanced carbon dioxide drawdown (Rayfuse et al. 2008, p. 307). Scientists have also examined the relationship between ocean fertilization and ocean acidification, concluding that, with fixed emissions of CO_2 into the atmosphere, ocean fertilization moderately mitigates changes in ocean carbonate chemistry near the ocean surface, but at the expense of further acidifying the deep ocean (Cao and Caldeira 2008). Other side effects include modification of the global macronutrient balance, modification of the global iron balance and changes to both pelagic and benthic ecosystem structures (Lampitt et al. 2008, pp. 3934–3938). The sustainability of ocean fertilization as a climate change mitigation option has also been called into question on the basis of the time frames and quantities of iron or other nutrients required for the process to be effective (Johnston et al. 1999, pp. 23–24; Rayfuse et al., 2008, p. 307). One projection estimates that approximately 470 000 tonnes of iron per year, spread over as much as 25 per cent of the ocean surface and repeated for an indefinite period would be needed for this method of carbon dioxide sequestration to be effective (Johnston et al. 1999, pp. 23–24; Rayfuse et al. 2008, p. 307).

In 2007, the IPCC concluded that ocean fertilization was 'speculative, unproven and with risks of unknown side effects'. Nevertheless, notwithstanding the scientific uncertainties and environmental concerns, a number of commercial ventures have shown interest in the process in recent years, hoping to reap financial benefits through the sale of carbon offsets. These activities have also demonstrated the financial and political risks associated with the process. Commercial iron fertilization activities have been proposed for areas in waters beyond national jurisdiction, raising the concerns of nearby coastal States as to the potential for damage to waters under their jurisdiction. A common feature of proposals for other methods of fertilization appears to have been that developing countries or countries in transition with less well-established environmental impact assessment (EIA) policies and processes have been targeted as conduits for the trials, thereby also raising the concerns of the international community.

In May 2007, United States-based company Planktos Corp announced plans to dissolve 100 tonnes of iron over a 10000 km^2 tract of high seas off the Galapagos Islands. This was to be the first of six large-scale pilot projects conducted by Planktos from 2007 to 2009 in the Pacific and Atlantic Oceans, each one lasting approximately four months. As a result of protests from the Ecuadorian Government, environmental NGOs and others, Planktos revised its plans, and its vessel eventually set sail in November 2007 for an unknown destination. In December 2007, the vessel was denied port entry by Spanish authorities in the Canary Islands and Planktos announced that it would be winding down its business ('Planktos shareholder update' 2007). Similarly, United States-based company, Climos, which has been active in promoting the commercial use of iron ocean fertilization as a climate change mitigation strategy, has recently ceased to conduct business activities (Climos 2011). In 2007, a proposed injection by Australian-based Ocean Nourishment Corp of 500 tonnes of urea into the Sulu Sea between the Philippines and Borneo through a pipeline from the Philippines coast was called off after it attracted criticism from local communities as well as regional and global environmental organizations concerned that adequate assessment of its impacts on the marine environment and marine biodiversity had not been undertaken (ETC Group 2007; Rayfuse et al. 2008, p. 299; Salleh 2007). More recent plans to pump urea into the Tasman Sea have been rejected by Australia and New Zealand. However, the company is pursuing possibilities with other Asian governments.

In February 2011, the In situ Iron Studies Consortium was established 'to resolve the impact of iron fertilization on marine ecosystems, to quantify its potential for removal of atmospheric carbon dioxide, and to improve our collective understanding of the changing ocean' (Isis Consortium 2011). The consortium consists of 13 institutions whose scientists are 'motivated to answer the unknowns regarding the role of iron in regulating the ocean's capacity to remove atmospheric carbon dioxide' (UN 1982). Consortium members have signed a Memorandum of Understanding that recommends support of open, peer-reviewed research to be conducted in accordance with international standards.

3. GENERAL PRINCIPLES OF INTERNATIONAL LAW APPLICABLE TO MITIGATION ACTIVITIES IN THE OCEANS

3.1 1982 LOSC Provisions

The basic legal framework for achieving protection and preservation of the marine environment is set out in Part XII of the 1982 United Nations Convention on the Law of the Sea (LOSC) (UN 1982). At the zenith of Part XII is Article 192, which codifies the general customary international law obligations on States Parties to ensure that activities under their jurisdiction or control do not cause harm to the environment of other States or to areas beyond national jurisdiction. Article 194(1) gives content to this general obligation by codifying the duty of States to prevent, reduce and control pollution of the marine environment from any source (Boyle 1992, p. 80). The global scope of this responsibility is manifest in Article 194(2), which refers to States Parties' duty to ensure that pollution arising from incidents or activities under their jurisdiction or control does not spread beyond the areas over which they exercise sovereign rights. An indicative list of the sources of marine pollution is contained in Article 194(3), which provides that States Parties shall take measures designed to minimize their effects to the fullest possible extent. Particularly relevant in the climate mitigation context, States are obliged to control the release of toxic, harmful or noxious substances, especially those that are persistent, from land-based sources, from or through the atmosphere or by dumping and pollution from other installations and devices operating in the marine environment.

In addition to these general duties to take measures to prevent, reduce and control marine pollution, Article 194(5) imposes a positive duty on States to take measures to protect and preserve rare and fragile ecosystems and the habitat of depleted, threatened or endangered species from marine pollution. This represents an early recognition of the need for ecosystem-based management of the oceans. The requirement in Article 195 that States not transfer, directly or indirectly, damage or hazards from one area to another has particular relevance to marine areas beyond national jurisdiction, as these areas have often been used as dumping grounds for vessel-source and other forms of pollution.

The practical issues of EIA and monitoring of the risks and effects of marine pollution in all areas of the sea are addressed in Articles 204 and 206, which require States Parties to assess, as far as practicable, the potential effects of planned activities under their control, which may cause substantial pollution or significant and harmful changes to the marine environment, and to communicate reports of the results of such assessments by publishing them or providing them to the competent international organizations.

Rather than being prescriptive in character, Part XII of the LOSC recognizes the role of competent international organizations such as the IMO and diplomatic conferences in supplementing the LOSC framework on marine pollution control with regulatory instruments that address specific forms of marine pollution. Article 197 provides for a duty on the part of States Parties to cooperate on a global and, as appropriate, regional basis, in the protection of the marine environment, directly or through competent international organizations, in formulating and elaborating international rules, standards and recommended practices and procedures for the protection and preservation of the marine environment. States must also cooperate directly or through competent international organizations for the purpose of promoting studies, undertaking programmes of scientific research and encouraging the exchange of information and data acquired about pollution of the marine environment, and to participate actively in programmes to assess the nature and extent of marine pollution, exposure to it and its pathways, risks and remedies.

Where climate change mitigation activities are experimental in character, the LOSC provisions on marine scientific research will apply to their conduct. Article 87 confirms the freedom of scientific research in high-seas areas, subject to the provisions of Part VI on the Continental Shelf and Part XIII on Marine Scientific Research. Articles 256 and 257 in Part XIII reinforce this freedom, providing that all States and competent international organizations have the right in conformity with the LOSC to conduct marine scientific research in the area and the water column beyond the limits of the exclusive economic zone. However, marine scientific research activities are expressly subject to the marine environmental protection provisions of the LOSC under Article 240(d) (Verlaan 2007, p. 211).

In addition, where climate change mitigation activities are conducted in high-seas areas above an area of extended continental shelf, States Parties and flag vessels under their jurisdiction or control must also have due regard for the sovereign rights of the relevant coastal State in relation to the living and non-living resources of the shelf. For example, Article 79(2) of the LOSC provides that although a coastal State may not impede the laying or maintenance of pipelines on the continental shelf beyond its territorial sea, it has the right to take reasonable measures for the prevention, reduction and control of pollution from such pipelines.

Implementation of these principles in areas under national jurisdiction falls under the control of the coastal State. However, in areas beyond national jurisdiction, implementation of these principles is largely dependent on flag State responsibility. Under Article 217 of the LOSC, flag States must ensure compliance by vessels flying their flag, with applicable international rules and standards established through the competent international organization and with their own laws and regulations for the prevention, reduction and control of marine pollution from vessels, including pollution by dumping (Birnie and Boyle 2002, p. 370; Molenaar 1998, p. 99). Flag States must provide for the effective enforcement of such rules, standards, laws and regulations, irrespective of where a violation occurs. The system of flag State jurisdiction over all forms of vessel-source pollution depends on the commitment and resources of States to monitor the compliance of their own fleets and take enforcement measures against delinquent vessels.

3.2 Complementary Principles for Regulating the Marine Environment

Since the 1972 Stockholm Declaration on the Human Environment (UN 1972), a number of other conservation principles have emerged that also apply to the protection of the marine environment, both within and beyond national jurisdiction (Verlaan 2007, pp. 210–211). Although generally expressed as being consistent with the provisions of the LOSC, and qualified with the prescription that they must be implemented consistently with the rights and obligations of States under the LOSC, these principles have followed a separate development trajectory. Thus, climate change mitigation activities in ocean areas are also subject to the evolving body of marine environmental protection principles; in particular, those articulated in the Rio Declaration (UN 1992) and the Convention on Biological Diversity (CBD) (1992).

3.2.1 The Rio Declaration and Agenda 21 Oceans chapter

The Rio Declaration, which grew out of the United Nations Conference on Environment and Development (UNCED) process (Freestone 1994, p. 216) sets out a number of principles applicable to the preservation and protection of the marine environment. One notable inclusion is that of the precautionary principle. Principle 15 of the Rio Declaration specifies that where there are threats of serious or irreversible damage to the environment, lack of full scientific certainty shall not be used as a reason for postponing cost effective measures to prevent environmental degradation. Application of the precautionary principle has particular relevance for marine areas beyond national jurisdiction because of the still developing state of scientific knowledge in relation to most aspects of the deep-sea environment and the wide array of new and emerging uses of these areas. This embryonic state of knowledge arguably imposes an even greater responsibility on the international community to adopt preventive strategies to protect this part of the global environment, as evidenced, for example in the number of global instruments in which the burden of proof is reversed, making it impermissible to conduct an activity in areas beyond national jurisdiction unless it can be shown that it will not cause unacceptable harm to the environment (Birnie and Boyle 2002, p. 118). The use of EIA processes for proposed activities that are likely to have a significant adverse impact on the environment is also encouraged in Principle 17 of the Rio Declaration.

Many of the principles contained in the Rio Declaration, including the precautionary approach and the recommendation that States conduct EIAs for proposed activities, are further elaborated in the UNCED Action Plan, Agenda 21 (UNCED 1993). Of particular relevance here, the Oceans chapter, Chapter 17, recognizes the underlying unity of the oceans, describing the oceans and all seas and adjacent coastal areas as 'an integrated whole that is an essential component of the global life support system' (UNCED 1993, para. 17.1). The primacy of the LOSC as the governing framework for the protection and sustainable development of the marine and coastal environment and its resources is acknowledged, but the need for fresh approaches to marine and coastal management that are 'integrated in content' and 'precautionary and anticipatory' in ambit is clearly stated (UNCED 1993, para. 17.1). The general obligation of States to protect and preserve the marine environment in Article 192 of the LOSC is expanded upon by specification of a set of objectives to guide States' efforts in arresting the degradation of the marine environment. Many of these are derived from the principles contained in the Rio Declaration. They include the application of preventive, precautionary and anticipatory approaches to reduce the risk of long-term and irreversible damage to the marine environment, the prior assessment of activities that may have significant adverse impacts on the environment, the integration of marine environmental protection considerations into social and economic development policies, incentives such as the 'polluter pays' principle to encourage the application of clean technologies, and other means consistent with the internalization of environmental costs (UNCED 1993, para. 17.22(a–d)).

3.2.2 1992 Convention on Biological Diversity

The provisions of the CBD are closely linked to the vision expounded in the Rio Declaration and Agenda 21 of integrated and ecosystem-based management of the environment, including of marine areas beyond national jurisdiction (Grubb et al. 1993, pp. 75–76). Biological diversity is an all-encompassing term, defined in Article 2 of the 1992 CBD as 'the variability among living organisms from all sources including, *inter alia*, terrestrial, marine and other aquatic ecosystems and the ecological complexes of which they are part', and including 'diversity within species, between species and ecosystems'. The obligation to conserve biodiversity contained in the CBD requires protection of a range of interlinked components in the marine environment, including species, habitats, ecosystems and genetic material, and takes into account the social, economic and political factors affecting the various components of marine biodiversity (Grubb et al. 1993, pp. 644, 646). Under Article 14 of the CBD, Contracting Parties must introduce EIA procedures for proposed projects that are likely to have significant adverse effects on biodiversity to avoid or minimize such effects. In the case of biological diversity beyond national jurisdiction, Article 5 of the CBD limits the obligations of Contracting Parties to a duty to cooperate in its conservation and sustainable use directly or through competent international organizations. Arguably, the obligations of States to protect and preserve the marine environment under Part XII of the LOSC must now be interpreted in the light of the provisions of the CBD (Birnie 1997, p. 338; Kimball 1995, pp. 769–771).

4. THE EMERGING LEGAL AND POLICY FRAMEWORK FOR MITIGATION ACTIVITIES IN THE OCEANS

4.1 Regulating Offshore Carbon Capture and Storage

4.1.1 Developments at the global level

As noted above, the LOSC requires States to take, individually and jointly, all measures necessary to prevent, reduce and control pollution of the marine environment, to prohibit the transfer, either directly or indirectly, of damage or hazards from one area to another, and to prohibit the transformation of one type of pollution to another (UN 1982, arts 192–196). Pollution is defined as:

> the introduction by man, directly or indirectly, of substances or energy into the marine environment, including estuaries, which results or is likely to result in such deleterious effects as harm to living resources and marine life, hazards to human health, hindrance to marine activities, including fishing and other legitimate uses of the seas, impairment of quality for use of sea water and reduction of amenities. (UN 1982, art 1(4))

It is not the nature of the substance, *per se*, that matters, but rather its potential for deleterious effects. Neither is the nature or purpose of the polluting activity relevant. States are to prevent, reduce and control pollution from all sources, whether generated from scientific research or from commercial operations, and whether transmitted from land-based sources, through the atmosphere, or from vessels, including from 'dumping' (UN 1982, art 196).

'Dumping' is defined in the LOSC as 'any deliberate disposal of wastes or other matter from vessels, aircraft, platforms or other man-made structures at sea' (IMO 1972, art 1; IMO 1996, art 1; UN 1982, art 1(5)). All States are required to adopt national laws to prevent and regulate dumping, and these laws must be no less effective than internationally agreed upon global rules and standards (UN 1982, art 210). These rules and standards are found in the 1972 Convention on the Prevention of Marine Pollution by Dumping of Wastes and other Matter (London Convention or LC) and the 1996 Protocol to the London Convention (London Protocol or LP). The LC applies to 'any deliberate disposal at sea of waste or other matter from vessels, aircraft, platforms or man-made structures at sea' (IMO 1972, art III 1(a)). Dumping of certain listed materials is wholly prohibited, while dumping of non-prohibited substances is only allowed subject to the requirements of prior EIA, permitting and on-going monitoring as set out in Annex III of the Convention. The LP was negotiated in 1996 to replace the LC. Although it has entered into force, it has limited participation (38 States Parties) and the two regimes operate in parallel. A fundamental premise of the LP is that Contracting Parties should avoid using the sea for the dumping of wastes and that any exceptional dumping of wastes at sea should be subject to rigorous risk assessment and control and scientifically based procedures for disposal (IMO 1996, art 2). Dumping of any waste or other matter is therefore strictly prohibited under the LP, except for those few materials specifically listed in Annex I (IMO 1996, art 4).

The LC makes no mention of the seabed or sub-seabed. Therefore, it is generally considered that the LC does not apply to OCCS. However, when the LP was negotiated, the definition of dumping was expanded to prohibit 'any storage of wastes or other matter in the seabed and the subsoil thereof from vessels, aircraft, platforms or other man-made structures at sea' (IMO 1996, art 1(4)(3)), except for those materials specifically listed in Annex I (IMO 1996, art 4). Annex I includes 'inert, inorganic geological material' and 'organic material of natural origin'. It is unlikely that carbon dioxide would fall into either of these categories. Nevertheless, as already mentioned, the direct injection of CO_2 into subsurface geological formations has been used as part of EOR projects since the early 1970s (GCCSI 2009; IPCC 2007, p. 199). These operations are generally considered to fall under the exception to dumping as 'placement for a purpose other than the mere disposal thereof'. However, since the 1990s, the technology has also been used to capture and store CO_2 from natural gas processing operations and, as mentioned in the discussion of carbon sequestration above, in 1996, the Norwegian oil company Statoil opened the world's first commercial offshore CO_2 storage site. While EOR operations might fall under the exception, the deliberate disposal of excess atmospheric CO_2 into commercially operated sub-seabed sites does not.

With some States increasingly promoting the concept of CCS as a means of alleviating atmospheric CO_2 levels, the LP was amended at the First Meeting of the Contracting Parties immediately after it came into force in 2006 to permit storage of carbon dioxide under the seabed (IMO 2007c). As amended (and entering into force on 10 February 2007 for all Contracting Parties to the Protocol), the LP allows for the dumping of 'carbon dioxide streams from carbon dioxide capture processes for sequestration' where disposal is into a sub-seabed geological formation and the streams consist overwhelmingly of carbon dioxide (they may contain incidental associated substances derived from the source material and the capture and sequestration processes used), and no wastes or other matter are added for the purpose of disposing of those wastes or other matter. In June 2006, the Scientific Group adopted a Risk Assessment and Management Framework to provide for characterization of the risks posed by CO_2 sequestration on a site-specific basis and to enable the collection of all necessary information for developing a management strategy to 'address uncertainties and any residual risks' (IMO 2006, Annex 3). Specific Guidelines for Assessment of Carbon Dioxide Streams for Disposal into Sub-seabed Geological Formations were adopted by the Parties in November 2007 to supplement the Guidelines for the Assessment of Wastes or Other Matter that May be Considered for Dumping, adopted in 1997 (IMO 2007b).

The Specific Guidelines take a precautionary approach to the process requiring Contracting Parties, under whose jurisdiction or control such activities are conducted, to issue a permit for the disposal, subject to stringent conditions being fulfilled (IMO 2007b, s. 9). The chemical and physical properties of carbon dioxide streams proposed for sub-seabed sequestration must be rigorously analysed and alternative methods of land-based disposal appropriately considered (IMO 2007b, ss. 3, 4). In addition, permit applicants must provide a geological assessment of the proposed site, which includes information on its long-term storage integrity, potential migration and leakage pathways over time, potential effects on the marine environment of leakage of carbon dioxide, and possibilities for monitoring, mitigation and remediation if leakage occurs (IMO 2007b, s. 6.2). Permit applicants must also provide information on the biological features and uses of the proposed site, including the presence of vulnerable ecosystems, critical habitats, spawning, nursery and recruitment areas for fish, shipping lanes, migration routes, military exclusion zones and engineering uses of the sea, such as undersea cables and energy conversion (IMO 2007b, s. 6.6). Applications for permits are required to evaluate the potential effects of a leakage of the carbon dioxide stream on human health, living resources, amenities and other legitimate uses of the sea (IMO 2007b, s. 7.6). This evaluation leads to an impact hypothesis forming the basis for a monitoring programme aimed at ensuring effective management of the disposal site and triggering mitigation or remediation plans if necessary (IMO 2007b, s. 7.11). In May 2008, a special reporting format was adopted to ensure adequate and accurate reporting on CO_2 streams for disposal into sub-seabed formations (IMO 2008, Annex 8).

In October 2009, the LP was further amended to allow the export of CO_2 for OCCS purposes. As originally adopted, Article 6 prohibits the export of wastes or other matter to other countries for dumping or incineration at sea. As amended, Article 6 now allows the export of carbon dioxide streams for disposal in accordance with Annex 1, provided that an agreement or arrangement has been entered into by the countries concerned and that agreement or arrangement includes confirmation and allocation of permitting responsibilities between the exporting and receiving countries and, in the case of export to non-Contracting Parties, provisions are, at a minimum, equivalent to those contained in the LP. Use of the word 'export' rather than 'trans-boundary movement' excludes migration of CO_2 after injection, thereby ensuring responsibility continues for trans-boundary migration after injection, while the requirements for agreement with non-Contracting Parties ensure that the Contracting Parties cannot contract out of their obligations under the LP. Until the amendment comes into force, export of CO_2 streams for OCCS will continue to be prohibited under the LP. Nevertheless, a work plan was adopted by the Scientific Groups in October 2010 for review of the Specific Guidelines to establish guidelines for the export of CO_2 for disposal in anticipation of the amendment coming into force (IMO 2010b).

While these comprehensive guidelines have been designed to avert the potential risks of this form of waste disposal at sea, it must be remembered that they only apply to the limited number of States Parties to the London Protocol. Nevertheless, these States are obliged to enact domestic legislation consistent with their international obligations. It is therefore germane to examine developments in national jurisdictions.

4.1.2 Developments in Australia

Australia, as one of the world's largest coal producers and exporters has emerged as a leading proponent of CCS projects both onshore and offshore. The Commonwealth Government had already developed the *Australian Regulatory Guiding Principles for Carbon Dioxide and Geological Storage* (Guiding Principles) in 2005, prior to the adoption by the LP of the Specific Guidelines in 2007. For OCCS projects, the Guiding Principles led to the Offshore Petroleum Amendment (Greenhouse Gas Storage) Act 2008, which amended the Offshore Petroleum Act 2006. The consolidated Act is now entitled the Offshore Petroleum and Greenhouse Gas Storage Act 2006.

The Guiding Principles were designed to facilitate a nationally consistent approach to carbon dioxide capture and geological storage across Commonwealth and State Governments in Australia's Federal System. They cover the assessment and approvals process for CCS activities; access and property rights to CCS sites; protocols on transport of CCS streams; requirements for monitoring and verification of the quantity of gas captured and stored; the net abatement of carbon dioxide emissions and the identification of and accounting for leakage; liability and post-closure responsibilities; and financial issues associated with the taxation liabilities of CCS projects and post-closure liabilities.

The Guiding Principles recognize the need for the Australian regulatory framework for CCS activities to be consistent with Australia's obligations under international law and to safeguard public interest, particularly to minimize risks to health, safety, environment, economic consequences and government accountabilities. These Principles are based on the Ecologically Sustainable Development Principles accepted by the Council of Australian Governments (COAG), which in turn reflect established international law principles on environmental protection. For example, COAG's agreed upon Ecologically Sustainable Development Principles are influenced by the precautionary principle, intergenerational equity and the 'polluter pays' principle, as well as by the need for decisions and actions on CCS activities to provide for broad community involvement on issues that affect those communities. The Guiding Principles recognize the applicability of the LC, its 1996 Protocol and the LOSC to OCCS activities, as well as the customary international law duty not to cause trans-boundary environmental damage; although no recognition is made in the Principles of the need not to cause trans-boundary damage to areas beyond national jurisdiction.

The need to subject CCS activity proposals to EIA processes prior to approval of a project is fully acknowledged in the Guiding Principles. The Principles recommend that existing legislation and regulations relating to CCS activities be modified and augmented where necessary to provide for consistent assessment and approval processes across all parts of a CCS project. For OCCS projects, this approach is consistent with Article 206 of the LOSC and Article 3 of the CBD, which require States Parties to conduct prior EIAs of activities or projects with the potential for significant adverse impacts on the marine environment and its biodiversity. In Australia's case, a proposed OCCS activity in an area beyond three nautical miles from the territorial sea baseline would be subject to an EIA process under the Environment Protection Biodiversity Conservation Act 1999 (Cth). At the stage when the carbon dioxide is being sequestered, the activity would also require a permit under the Environment Protection (Sea Dumping) Act 1981 (Cth), which implements Australia's obligations under Annex 2 of the LP relating to Waste Assessment.

In relation to access and property rights to greenhouse gas storage sites both onshore and offshore, the Guiding Principles recommend that surface and subsurface rights for CCS should provide certainty to rights holders of their entitlement and obligations and that these rights should be based on established legislative and regulatory arrangements, custom and practice. The Principles also recommend that additional governmental regulation be introduced to define property rights in relation to CCS. These recommendations were implemented in the 2008 amendments to the Offshore Petroleum and Greenhouse Gas Storage Act 2006. These amendments introduced a system of greenhouse gas assessment permits and greenhouse gas holding leases over blocks in Australia's offshore area, through which permit holders can explore the area for potential greenhouse gas storage sites and eventually be granted a greenhouse gas injection licence. These permits, leases and licences are granted by the Minister for Resources, Energy and Tourism and administered through the Department for Resources, Energy and Tourism. In approving grants for greenhouse gas assessment permits, the Minister must consider whether greenhouse gas storage operations in a particular offshore area would have a significant adverse impact on existing petroleum exploration or recovery operations.

The Guiding Principles emphasize the need for monitoring and verification of the stored CCS streams to ensure operationally safe performance and to minimize the risk of leakage into the surrounding environment. They specify the content of both the monitoring information and the transparency measures in relation to the information obtained. The regulatory framework should provide for the accuracy, quantity, composition and location of gas captured, transported, injected and stored and the net abatement of emissions. Regulation should also provide for appropriate monitoring and verification requirements, enabling the generation of clear comprehensive, timely, accurate and publicly accessible information that can be used to effectively and responsibly manage environmental, health, safety and economic risks. The Environmental Guidelines for Carbon Dioxide Capture and Geological Storage issued by the Environment Protection and Heritage Council in 2009 endorse these principles, specifying that CCS projects that become operational in the next decade must include comprehensive monitoring regimes including in-hole geo-chemical monitoring, geophysical, including seismic, monitoring, and modelling of the carbon dioxide plume. This will be an essential condition of approval for all CCS projects. For CCS projects in the Commonwealth marine area beyond three nautical miles, development of the monitoring and verification conditions is likely to be part of the EIA process and final project approval. Under the Offshore Petroleum and Greenhouse Gas Storage Act 2006, decommissioning reports must be submitted to the Minister for Resources, Energy and Tourism, together with suggestions for monitoring, measurement and verification. The requirement for a site-closing certificate includes a thorough assessment of the migratory behaviour of the injected greenhouse gas.

The Guiding Principles identify the need to define liabilities and post-closure responsibilities associated with CCS projects. The key liability foreshadowed in the Principles is the potential monetary compensation arising from a leakage of the CCS stream because of negligence during any part of the CCS process. During the currency of the CCS operation, the liability to pay monetary compensation would fall on the project operator, who has a duty to take reasonable care to avoid harm to a person or property, but may have failed to do so. The Principles also indicate the need to develop a policy for acceptance of post-closure responsibilities arising from risks associated with long-term storage of the carbon dioxide following the decommissioning of the storage site. These recommendations have been implemented in the 2008 amendments to the Offshore Petroleum and Greenhouse Gas Storage Act 2006 (Cth). The Act imposes statutory liability for damages on the holder of a greenhouse gas injection licence until a site-closing certificate has been issued. It also contains a provision on transfer of long-term liability from the operator to the Government at the end of a closure assurance period, which is to be a minimum of five years. The Act includes a 20-year sunset period on the proponent's liability for damages.

The evolving regulatory environment for offshore carbon capture projects in Australia is defined by its complex and fragmentary character. The commercial and environmental viability of future CCS activities in Australia's offshore areas will depend on strong collaborative governance and adaptive management across Commonwealth and State jurisdictions and relevant industry sectors. Within the Asia-Pacific region, Australia is the first country to pass dedicated carbon capture and storage legislation, although such activities in other Asia-Pacific jurisdictions will undoubtedly fall within established regulatory frameworks for EIA, planning, pollution prevention and waste disposal.

4.2 Regulating Ocean Fertilization

The long-term environmental impacts of ocean fertilization are still uncertain and the regulatory framework for this process is still developing. In part, development of a single coherent regulatory approach is complicated by the range of proposed and actual fertilization techniques. Different legal considerations arise depending on the technique used (for example, ocean-based fertilization, land-based fertilization or wave-mixing machines suspended in the water column) and the locus of the fertilization (that is, whether fertilization activities occur in areas beyond national jurisdiction or in areas within national jurisdiction) (Rayfuse 2008, p. 920). In addition, this complex matrix of legal relationships and regulatory possibilities may be further complicated when the purpose of the fertilization is considered. This is because fertilization for climate mitigation purposes might, arguably, be distinguishable from fertilization for ocean nourishment and fish propagation purposes. While ocean fertilization activities conducted in marine areas within national jurisdiction will be subject to coastal State control, this jurisdiction must be exercised consistent with international obligations. It is thus appropriate to examine the emerging global framework for the regulation of ocean fertilization activities.

Like OCCS, it is in the context of the international regime for the control of dumping that debates over regulation of ocean fertilization have received most attention. As noted above, for States Parties to the LC, dumping of non-prohibited substances is only allowed subject to the requirements of prior EIA, permitting and on-going monitoring as set out in Annex III of the Convention. For parties to the LP, the dumping of all waste and other matter is prohibited – with the exception of the five listed categories of substances, the dumping of which is nevertheless subject to the stringent assessment, permitting and on-going monitoring requirements of Annex 2 of the Protocol. None of the fertilizers proposed for use in ocean fertilization fall into any of these categories (Freestone and Rayfuse 2008, pp. 227–233; Rayfuse et al. 2008, p. 307). In other words, the use of these 'fertilizers' is *prima facie* banned.

The central issue for ocean fertilization is whether it is exempt from the ban on dumping by virtue of the operation of the exception to the definition of dumping found in the LOSC, LC and LP (Freestone and

Rayfuse 2008, pp. 307–317). Stated in the same terms in each Convention, dumping is defined as *not* including 'placement of matter for a purpose other than the mere disposal thereof, provided that such placement is not contrary to the aims of' the relevant Convention. This qualification on the definition of dumping potentially excludes ocean fertilization from the general prohibition on dumping if the fertilization is for the purpose of climate mitigation or other commercial and environmental purposes, such as fisheries enhancement. However, in view of its potentially adverse effects on the marine environment, even the experimental phases of ocean fertilization may be regarded as contrary to the marine environmental protection aims of the LOSC, the LC and/or the LP (Rayfuse et al. 2008, pp. 313–315). If this is the case then, for States Parties to the LC, ocean fertilization activities will be subject to the permitting requirements set out in the Convention. However, for States Parties to the more stringent LP, ocean fertilization is prohibited.

In May 2007, the parties to the LC and LP were confronted for the first time with proposals for large-scale commercial ocean iron fertilization projects. As noted in Section 2.2 above, a number of United States and Australian companies were promoting ocean fertilization as a tool to buffer ocean acidity, replenish the marine food chain and sequester CO_2, while inviting investors and green co-sponsors to finance their activities in return for the provision of carbon credits to offset investors' CO_2 emissions. A 'statement of concern' adopted by the Scientific Groups of the LC/LP in July 2007 'noted with concern the potential for [ocean fertilization activities] to have negative impacts on the marine environment and human health' and recommended that the parties to the LC and LP consider the issue with a view to its regulation (Scientific Groups of the London Convention and the London Protocol 2007). This statement of concern was endorsed by the States Parties during their joint annual meeting in November 2007, at which the parties agreed that while it was within the purview of each State to consider proposals for ocean fertilization projects on a case-by-case basis in accordance with the Convention and/or Protocol, knowledge about the effectiveness and potential environmental impacts of open-ocean fertilization was currently insufficient to justify large-scale projects. They also agreed that ocean fertilization falls within their regulatory competence and that they would 'further study this issue from scientific and legal perspectives with a view to its regulation' (IMO 2007b).

In May 2008, the Scientific Group reviewed the evidence on open-ocean fertilization and concluded that 'based on scientific projections, there is the potential for significant risks of harm to the marine environment', even if direct scientific evidence on the environmental impact was still lacking.

This decision prompted the Conference of the Parties of the CBD, at their ninth meeting in May 2008, to request parties and urge other governments:

> in accordance with the precautionary approach to ensure that ocean fertilization activities do not take place until there is an adequate scientific basis on which to justify such activities, including assessing associated risks, and a global transparent and effective control and regulatory mechanism is in place for those activities; with the exception of small scale scientific research within national jurisdiction. (COP-9 2008, s. C)

An exception was noted in the case of 'small scale scientific research studies within coastal waters', which:

> should only be authorised if justified by the need to gather specific scientific data, and should also be subject to a thorough prior assessment of the potential impacts of the research studies on the marine environment, and be strictly controlled, and not be used for generating and selling carbon offsets or any other commercial purposes. (COP-9 2008, s. C)

In October 2008, the parties to the LC/LP adopted a non-binding resolution agreeing that:

> given the present state of knowledge ocean fertilization activities other than legitimate scientific research should not be allowed . . . [and that] ocean fertilisation activities other than legitimate scientific research, should be considered as contrary to the aims of the Convention and Protocol and not currently qualify for any exemption from the definition of dumping. (IMO 2008, paras 4.1–4.18, Annexes 2, 5)

They identified the need for preparatory work on technical and scientific issues and agreed to consider further a potentially legally binding resolution or an amendment to the LP at their 2009 session. An intersessional Technical Working Group on Ocean Fertilization was established to develop an Assessment Framework for Scientific Research Involving Ocean Fertilization to provide a mechanism for assessing, on a case-by-case basis, whether proposals for ocean fertilization activities represent legitimate scientific research (IMO 2008, para. 2.3). The draft Assessment Framework (IMO 2010a, Annex 2) was reviewed by the Scientific Groups in June 2009 and adopted as a 'work in progress' (IMO 2010a, paras 2.18–2.29). The draft was tabled again during an extraordinary session of the Scientific Groups in October 2010 (IMO 2010a, Annex 5), at which time further revisions were made and it was adopted, by consensus, in a non-binding resolution at the October 2010 meeting of the parties (IMO 2010b, Annex 5).

The Assessment Framework (AF) describes itself as a 'tool . . . to determine if the proposed activity constitutes legitimate scientific research that is not contrary to the [LC/LP] aims'. It sets out a two-stage process involving an Initial Assessment and an Environmental Assessment. The purpose of the Initial Assessment is to determine whether the proposed ocean fertilization activity constitutes legitimate scientific research. To qualify as such, the proposed activity must have 'proper scientific attributes', which means:

1. The proposed activity should be designed to answer questions that will add to the body of scientific knowledge. Proposals should state their rationale, research goals, scientific hypotheses and methods, scale, timings and locations with clear justification for why the expected outcomes cannot reasonably be achieved by other methods.
2. Economic interests should not influence the design, conduct and/or outcomes of the proposed activity. There should not be any financial and/or economic gain arising directly from the experiment or its outcomes. This should not preclude payment for services rendered in support of the experiment of the future financial impacts of patented technology.
3. The proposed activity should be subject to scientific peer-review at appropriate stages in the assessment process. The outcomes of the scientific peer review should be taken into consideration by the Contracting Parties. The peer-review methodology should be stated and the outcomes of the peer review of successful proposals should be made publicly available together with the details of the project.
4. The proponents of the proposed activity should make a commitment to publish the results in peer-reviewed scientific publications and include a plan in the proposal to make the data and outcomes publicly available over a specified period.

Proposals that meet these criteria may then proceed to the next stage, the Environmental Assessment, which includes requirements of risk management and monitoring. The Environmental Assessment stage entails a number of components, including the problem formulation, a site selection and description, an exposure assessment, an effects assessment, risk characterization and risk management sections (IMO 2010b, Annex 6). Only after completion of the Environmental Assessment is it decided whether the proposed activity constitutes legitimate scientific research that is not contrary to the aims of the LC/LP, and whether it should thus be permitted to proceed.

Importantly, every experiment, regardless of size or scale, is to be assessed in accordance with the AF. This is fully consistent with the

LOSC, which requires all activities affecting the marine environment to comply with its marine environmental provisions (Verlaan 2007, p. 216). While it is acknowledged that the information requirements will vary according to the nature of each experiment, it would be incompatible with the AF for parties to establish their own national thresholds to exempt some experiments. The AF is thus also consistent with, and possibly more stringent than, the CBD moratorium.

The AF represents a significant achievement in providing an environmentally responsible mechanism to assess and control ocean fertilization activities. However, as noted above, the AF is non-binding. In 2009, an intersessional Working Group on Ocean Fertilization was established to 'focus on deepening the understanding of the implications of legally binding options to enable the informed consideration and discussion of the governing bodies' (IMO 2009).

Numerous options for a legally binding measure on ocean fertilization have been identified and these can be essentially divided along two lines. Some States support a legally binding interpretative resolution that considers ocean fertilization for legitimate scientific research purposes to be 'placement for a purpose other than the mere disposal thereof' and thus not dumping. Ocean fertilization for all other purposes would be contrary to the aims of the LC/LP, and thus dumping. The advantages of this approach are that it would not require an amendment to the LC/LP or their annexes; it would address both the parties to the LC and the LP; and it would be adaptable to regulation of other marine geo-engineering activities in the future. However, a key disadvantage is that any legitimate scientific research ocean fertilization activities that had been approved through the AF process would not be subject to the legally binding permitting regime required for approved dumping operations, including its consultation and reporting requirements. Other States support adoption of an amendment to some or all of Annex 1 of the LP, the definition of dumping, the exclusions for dumping, or inclusion of a new stand-alone article in the LP on ocean fertilization (IMO 2010a). Under these various proposals, ocean fertilization for legitimate scientific purposes would be permitted under the LP and subject to its permitting requirements. Australia and New Zealand in particular take the position that an amendment to Annex 1 of the LP specifically permitting legitimate scientific research ocean fertilization activities is necessary to ensure its effective regulation through the permitting regime. In effect, this would put such activities on the same footing as OCCS. Other States remain strongly opposed to any regulation of ocean fertilization whatsoever.

Of course, even adoption by the LC/LP parties of a legally binding option, whatever it might be, will not wholly guarantee the effective regulation and control of ocean fertilization activities, particularly where they take place in areas beyond national jurisdiction. The LC and LP are only binding on their parties. No matter how strict an approach they take, the very real potential exists for proponents of ocean fertilization to undermine the LC/LP regulatory efforts by conducting their activities through non-contracting parties. In this respect, given its near global adherence, the CBD moratorium on ocean fertilization represents a critically useful adjunct to the work of the LC/LP.

Indeed, the CBD may be seen as something of a catalyst to the LC/LP process. Having provided impetus to the LC/LP discussion with respect to ocean fertilization, the CBD may now also have set the agenda for further action on the broader issue of marine geo-engineering in general. In 2009, the parties to the LC/LP considered whether the scope for regulation should be widened to cover other emerging marine geo-engineering proposals, or whether regulation should remain focused only on ocean fertilization as a sub-set of marine geo-engineering. At the time, it was agreed to maintain the focus on ocean fertilization, but that an exploration of other marine geo-engineering activities and their possible impacts on the marine environment would be conducted in the future (IMO 2009, para. 4.20). In 2010, the parties to the CBD took a more decisive step by extending the moratorium on ocean fertilization to encompass all climate-related geo-engineering activities that may affect biodiversity (COP-10 2010). Clearly, whatever procedures are adopted by the parties to the LC/LP to deal with ocean fertilization, these will have resonance for the regulation of other marine geo-engineering activities in the future.

5. CONCLUSION

The alacrity with which the LP was amended to allow for OCCS has been criticized by commentators who consider the amendment to be both inconsistent with Article 6 of the LP and a fundamental violation of Article 195 of the LOSC, which prohibits the transfer of pollution from one area to another (Verlaan 2009, p. 457). A different regulatory path has been followed for ocean fertilization.

The risk-assessment process adopted for ocean fertilization under the LC/LP for legitimate and small-scale scientific experiments provides a model for States Parties to apply modern international environmental law principles to protect the marine environment and conserve biodiversity, particularly where experiments occur beyond national jurisdiction. However, this model has obvious limitations, as it is designed for a specific activity that falls within the regulatory scope of the LC/LP, only binding

States Parties to the LC/LP and relying on individual flag State responsibility for the implementation of the risk-assessment process. Other climate change mitigation activities involving the oceans may be subjected to similar examination in the future by the LC/LP Scientific Groups if such activities fall within the regulatory ambit of these conventions. However, the proliferation of such activities suggests the need for a more integrated system of global, sectoral and regional instruments to provide a more comprehensive system of prior EIA and on-going monitoring of the long-term effects of such activities on the receiving environment.

Arbitrary human intrusions into previously undisturbed marine domains have the potential to harm the intricate links between complex marine ecosystems and to erode components of marine biodiversity. Protection of the vast tracts of ocean from the adverse impacts of new and emerging uses such as climate change mitigation activities requires concentrated global, regional and national investment into coordinating and extending environmental protection regimes and developing assessment frameworks. Enhanced environmental protection for the oceans will require concerted action by the international community to put in place best practice guidelines and measures to assess and minimize the adverse impacts of emerging climate change mitigation activities on all areas of the ocean.

REFERENCES

Abanedes, Juan C., et al. (2005), *IPCC Special Report on Carbon Dioxide Capture and Storage*, Cambridge: Cambridge University Press.

Alfred Wegener Institute (2009), 'Lohafex', accessed 26th July 2012 at: http://www.awi.de/en/news/selected_news/2009/lohafex/.

Aumont, O. and L. Bopp (2006), 'Globalising results from ocean in situ iron fertilisation studies', *Global Biogeochemical Cycles*, **20** (2), 1–15.

Australian Regulatory Guiding Principles for Carbon Dioxide and Geological Storage (2005), Ministerial Council on Mineral and Petroleum Resources, accessed 26th July 2012 at: http://www.ret.gov.au/resources/Documents/ccs/CCS_Aust_Regulatory_Guiding_Principles.pdf.

Birnie, Patricia (1997), 'Are twentieth-century marine conservation conventions adaptable to twenty first century goals and principles? Part 1', *The International Journal of Marine and Coastal Law*, **12** (3) 307–339.

Birnie, Patricia and Alan Boyle (2002), *International Law and the Environment* (2nd edn), Oxford: Oxford University Press.

Boyd, Philip W., T. Jickells, C.S. Law, S. Blain, E.A. Boyle, K.O. Buesseler . . . A.J. Watson (2007),'Mesoscale iron enrichments experiments 1993–2005: synthesis and future directions', *Science*, **315** (5812), 612–617.

Boyle, Alan (1992), 'Protecting the marine environment: some problems and developments in the Law of the Sea', *Marine Policy*, **16** (2), 79–85.

Brewster Weeks, Ann (2007), 'Sub-seabed carbon dioxide sequestration as a climate mitigation option for the Eastern United States: a preliminary assessment of technology and law', *Ocean and Coastal Law Journal*, **12** (2), 245–268.

Buesseler, Ken O. and Philip W. Boyd (2003), 'Will ocean fertilisation work?' *Science*, **300** (5616), 67–68.

Cao, L. and K. Caldeira (2008), *Ocean Fertilization and Ocean Acidification*, American Geophysical Union, Fall Meeting, accessed 26th July 2012 at: http://adsabs.harvard.edu/abs/2008AGUFM.B22C..06C.

CBD (1992), *Convention on Biological Diversity*, opened for signature on 22 May 1992, entered into force 29 December 1993, 31 ILM 822.

Climos (2011), 'Climos', accessed 26th July 2012 at: http://www.climos.com/climos_notes.php.

COP-9 (2008), *COP 9 Decision XI/16 on Biodiversity and Climate Change*, presented at the Ninth Meeting of the States Parties to the Convention on Biological Diversity (COP 9), Bonn, 19–30 May, last accessed 27 August 2008 at: http://www.cbd.int/decisions/cop9/?m=COP-09&id=11659&lg=0.

COP-10 (2010), *Decision on Geoengineering*, presented at the Tenth Meeting of the States Parties to the Convention on Biological Diversity (COP 10), Nagoya, 18–29 October.

Cullen, John L. and Philip W. Boyd (2008), 'Predicting and verifying the intended and unintended consequences of large-scale ocean iron fertilization', *Marine Ecology Progress Series*, **364**, 213–218.

de Baar, H.J.W., Philip W. Boyd, Kenneth H. Coale, Michael R. Landry, Atsushi Tsuda, Philipp Assmy . . . Chi-Shing Wong (2005), 'Synthesis of iron fertilisation experiments: from the iron age to the age of enlightenment', *Journal of Geophysical Research – Oceans*, **110**, 1–24.

ETC Group (2007), 'Ocean nourishment corporation plans imminent urea dumping experiment in Southeast Asian seas. Civil society groups call on London Convention to halt marine dumping geo-engineering experiments', *ETC Group*, 5 November, last accessed 31 August 2008 at: http://rebelsoftheworld.blogspot.com.au/2008/01/ocean-nourishment-corporation-plans.html.

Freestone, David (1994), 'The road from Rio: international environmental law after the Earth Summit', *Journal of Environmental Law*, **6** (2), 193–218.

Freestone, David and Rosemary Rayfuse (2008), 'Ocean iron fertilisation and international law', *Marine Ecology Progress Series*, **364**, 227–233.

GCCSI (2009), *Strategic Analysis of the Global Status of Carbon Capture and Storage, Report 5: Synthesis Report*, accessed 26th July 2012 at: http://cdn.globalccsinstitute.com/sites/default/files/publications/5751/report-5-synthesis-report.pdf.

GCCSI (2010), 'About the Institute', accessed 26th July 2012 at: http://www.globalccsinstitute.com/institute.

Grubb, Michael, Matthias Koch, Koy Thomson, Abby Munson and Francis Sullivan (1993), *The 'Earth Summit' Agreements: A Guide and Assessment*, London: Earthscan Publications.

IEA (2008), *Energy Technology Perspectives*, International Energy Agency.

IMO (1972), *Convention on the Prevention of Marine Pollution by Dumping of Wastes and Other Matter*, opened for signature 29 December 1972, entered into force 30 August 1975, 1976 UKTS 43.

IMO (1996), *Protocol to the Convention on the Prevention of Marine Pollution by Dumping of Wastes and Other Matter,* opened for signature 7 November 1996, entered into force 24 March 2006, 36 ILM 1.

IMO (2006), *Report of the Meeting of the SG Intercessional Technical Working Group on CO_2 Sequestration, 3–7 April*, accessed 26th July 2012 at: http://www.shipadm.org/pages/8419/1-INF5.pdf.

IMO (2007a), 'CO2 sequestration: frequently asked questions', accessed 26th July 2012 at: http://www.imo.org/environment/mainframe.asp?topic_id=1548.

IMO (2007b), *Report of the Twenty-Ninth Consultative Meeting of the Contracting Parties to the Convention on the Prevention of Marine Pollution by Dumping of Wastes and other Matter 1972 and Second Meeting of the Contracting Parties to the 1996 Protocol Thereto*, IMO Doc. LC 29/17.

IMO (2007c), Press Briefing 5, 9 February 2007, last accessed 10 June 2008 at: http://www.imo.org/Safe ty/mainframe.asp?topic_id=1472&4doc_id=7772.

IMO (2008), *Report of the Thirtieth Consultative Meeting of Contracting Parties to the London Convention and Third Meeting of Contracting Parties to the London Protocol*, IMO Doc. LC 30/16, accessed 26th July 2012 at: http://www.ngo.grida.no/wwfneap/Publication/Submissions/OSPAR2009/WWF_BDC09_OF_Annex2.pdf.

IMO (2009), *Report of the Thirty-First Consultative Meeting of Contracting Parties to the London Convention and Fourth Meeting of Contracting Parties to the London Protocol*, IMO Doc. LC 31/15.

IMO (2010a), *Report of the Second Meeting of the LP Intersessional Group on Ocean Fertilization*, IMO Doc. LP CO2 3/7, accessed 26th July 2012 at: http://www.imo.org/blast/blastDataHelper.asp?data_id=30735&filename=7.pdf.

IMO (2010b), *Report of the Thirty-Second Consultative Meeting of Contracting Parties to the London Convention and Fifth Meeting of Contracting Parties to the London Protocol*, IMO Doc. LC 32/15, accessed 26th July 2012 at: http://www.ucl.ac.uk/cclp/pdf/Protocol15.pdf.

IPCC (2007), *Fourth Assessment Report*, accessed 26th July 2012 at: http://www.ipcc.ch/publications_and_data/publications_and_data_reports.shtml#.UBXj0GGICAg.

Isis Consortium (2011), 'What is the Isis Consortium', accessed 26th July 2012 at: http://www.isisconsortium.org/home.

Johnston, Paul, David Santillo and Ruth Stringer (1999), *Ocean Disposal/Sequestration of Carbon Dioxide from Fossil Fuel Production and Use: An Overview of Rationale, Techniques and Implications*, Greenpeace Research Laboratories Technical Note 01/99, Amsterdam: Greenpeace International.

Kimball, Lee A. (1995), 'The Biodiversity Convention: how to make it work', *Vanderbilt Journal of Transnational Law*, **28** (4), 703–754.

Koslow, Tony (2007), *The Silent Deep*, Sydney, NSW: UNSW Press.

Lampitt, R.S., E.P. Achterberg, T.R. Anderson, J.A. Hughes, M.D. Iglesias-Rodriguez, B.A. Kelly-Gerreyn . . . A. Yool (2008), 'Ocean fertilisation: a potential means of geoengineering?' *Philosophical Transactions of the Royal Society A: Mathematical, Physical and Engineering Sciences*, **366** (1882), 3919–3945.

Lin, Albert C. (2009), 'Geoengineering Governance', *Issues in Legal Scholarship*, **8** (3), 1–24.

Molenaar, Erik J. (1998), *Coastal State Jurisdiction over Vessel Source Pollution*, The Hague: Kluwer Law International.

'Planktos shareholder update' (2007), *Business Wire*, 19 December, accessed 26th July 2012 at: http://newenergytimes.com/v2/inthenews/2007/Q4/BW-PlanktosShareholderUpdate.shtml.

Preston, B.L. and R.N. Jones (2006), *Climate Change Impacts on Australia and*

the Benefits of Early Action to Reduce Global Greenhouse Gas Emissions: A Consultancy Report for the Australian Business Roundtable on Climate Change, Canberra, ACT: CSIRO.
Rayfuse, Rosemary (2008), 'Drowning our sorrows to create a carbon free future? Some international legal considerations relating to sequestering carbon by fertilising the oceans' *UNSW Law Journal*, **31**, 919–930.
Rayfuse, Rosemary, Mark G. Lawrence and Kristina M. Gjerde (2008), 'Ocean fertilisation and climate change: the need to regulate emerging high seas uses', *The International Journal of Marine and Coastal Law*, **23** (2), 297–326.
Salleh, Anna (2007), 'Urea "climate solution" may backfire', *ABC Science Online*, 9 November, accessed 26th July 2012 at: http://www.abc.net.au/science/articles/2007/11/09/2085584.htm.
Schneider, Stephen (2008), 'Geoengineering: could we or should we make it work?' *Philosophical Transactions of the Royal Society A: Mathematical, Physical and Engineering Sciences*, **366** (1882), 3843–3862.
Scientific Group of the London Protocol (2010), *Marine Geo-engineering – Types of Schemes Proposed to Date*, Submitted by the Chairman of the Scientific Groups to the Thirty-Third Meeting of the Scientific Group of the London Convention and the Fourth Meeting of the Scientific Group of the London Protocol, IMO Doc. LC/SG 33/13.
Scientific Groups of the London Convention and the London Protocol (2007), *Statement of Concern Regarding Iron Fertilisation of the Ocean to Sequester CO2*, IMO Doc. LC-LP.1/Circ.14.
Scott, K. (2005), 'The day after tomorrow: ocean CO_2 sequestration and the future of climate change', *Georgetown International Environmental Law Review*, **18**, 57–108.
Stern, Nicholas et al. (2006), *Stern Review: The Economics of Climate Change*, London: HM Treasury.
UN (1972), *Declaration of the United Nations Conference on the Human Environment*, Stockholm, 16 June 1972, (1972) 11 ILM 1416.
UN (1982), *United Nations Convention on the Law of the Sea*, opened for signature on 10 December 1982, entered into force 16 November 1994, 1833 UNTS 3.
UN (1992), *Rio Declaration on Environment and Development*, UN Doc. A/CONF.151/26 (Vol. 1), accessed 7th August 2012 at: http://www.un.org/documents/ga/conf151/aconf15126-1annex1.htm.
UNCED (1993), United Nations Conference on Environment and Development, Agenda 21: Chapter 17, UNCED Report, A/CONF.151/26/Rev.1 (vol.1).
Verlaan, P. (2007), 'Experimental activities that intentionally perturb the marine environment: implications for marine environmental protection and marine scientific research provisions of the 1982 United Nations Convention on the Law of the Sea', *Marine Policy*, **31** (2), 210–216.
Verlaan, P. (2009), 'Geoengineering, the Law of the Sea and climate change', *Carbon and Climate Law Review*, **3** (4), 446–458.

Index